27304-14

Reinforcing Concrete

OVERVIEW

In any structure built with concrete, the foundation and any concrete walls and floors will be reinforced. Failure to use the correct amount and type of reinforcing material may result in a catastrophic failure of the structure. The majority of reinforcing material is steel bars that are inserted into the form before the concrete is placed. It is critical to use the correct reinforcing material and supports, place them in accordance with the project drawings, and use the correct methods to tie the reinforcing material. While this process may seem simple on the surface, it requires training and experience in order to do it correctly.

Module Five

Trainees with successful module completions may be eligible for credentialing through NCCER's National Registry. To learn more, go to **www.nccer.org** or contact us at **1.888.622.3720**. Our website has information on the latest product releases and training, as well as online versions of our *Cornerstone* magazine and Pearson's product catalog.

Your feedback is welcome. You may email your comments to **curriculum@nccer.org,** send general comments and inquiries to **info@nccer.org**, or fill in the User Update form at the back of this module.

This information is general in nature and intended for training purposes only. Actual performance of activities described in this manual requires compliance with all applicable operating, service, maintenance, and safety procedures under the direction of qualified personnel. References in this manual to patented or proprietary devices do not constitute a recommendation of their use.

27304-14
REINFORCING CONCRETE

Objectives

When you have completed this module, you will be able to do the following:

1. List applications of reinforced concrete.
   a. Describe how forces are resisted in concrete through the use of reinforcing bars.
   b. List applications for reinforced structural concrete.
   c. Discuss how posttensioned concrete is created.
2. Describe the general requirements for working with reinforcing steel, including tools, equipment, and fabricating methods.
   a. List general safety precautions when working with reinforcing steel.
   b. Describe the general characteristics of reinforcing steel.
   c. Discuss how reinforcing steel is fabricated.
   d. Explain the purpose of bar supports.
   e. Explain how welded-wire fabric reinforcement is used to reinforce concrete.
3. Describe methods by which reinforcing bars may be bent and cut in the field.
   a. Describe how to cut rebar.
   b. Describe how to bend rebar.
4. Explain the methods for placing reinforcing steel.
   a. Discuss the proper method for tying and splicing reinforcing steel.
   b. Explain the proper procedure for placing reinforcing steel.

Performance Tasks

Under the supervision of your instructor, you should be able to do the following:

1. Use appropriate tools to cut and bend reinforcing bars.
2. Demonstrate five types of ties for reinforcing bars.
3. Demonstrate proper lap splicing of reinforcing bars using wire ties.
4. Demonstrate the proper placement, spacing, tying, and support for reinforcing bars.

Trade Terms

Abutment	Dowel	Schedule
Band	Far face	Simple beam
Bar list	Flat slab	Single-curtain wall
Beam	Girder	Sleeve
Bent	Hickey bar	Span
Bundle of bars	Hook	Staggered splices
Caissons	Inserts	Stirrups
Column	Lapped splice	Strips
Column spirals	Near face	Support bars
Column ties	Pile cap	Temperature bars
Concrete cover	Pitch	Template
Contact splice	Placing drawings	Tie
Continuous beam	Rebar horses	Tie wire
CRSI	Reinforced concrete	Weephole
Double-curtain wall	Retaining wall	

Industry-Recognized Credentials

If you're training through an NCCER-accredited sponsor, you may be eligible for credentials from NCCER's Registry. The ID number for this module is 27304-14. Note that this module may have been used in other NCCER curricula and may apply to other level completions. Contact NCCER's Registry at 888.622.3720 or go to **www.nccer.org** for more information.

Code Note

Codes vary among jurisdictions. Because of the variations in code, consult the applicable code whenever regulations are in question. Referring to an incorrect set of codes can cause as much trouble as failing to reference codes altogether. Obtain, review, and familiarize yourself with your local adopted code.

Contents

Topics to be presented in this module include:

Figures and Tables

SECTION ONE

1.0.0 REINFORCED STEEL APPLICATIONS

Objective

List applications of reinforced concrete.
  a. Describe how forces are resisted in concrete through the use of reinforcing bars.
  b. List applications for reinforced structural concrete.
  c. Discuss how posttensioned concrete is created.

Trade Terms

Abutment: The supporting substructure at each end of a bridge.

Beam: A horizontal structural member.

Bent: A self-supporting frame having at least two legs and placed at right angles to the length of the structure it supports, such as the columns and cap supporting the spans of a bridge.

Caissons: Piers usually extending through water or soft soil to solid earth or rock; also refers to cast-in-place, drilled-hole piles.

Column: A post or vertical structural member supporting a floor beam, girder, or other horizontal member and carrying a primarily vertical load.

Column spirals: Columns in which the vertical bars are enclosed within a spiral that functions like a column tie.

Column ties: Bars that are bent into square, rectangular, U-shaped, circular, or other shapes for the purpose of holding vertical column bars laterally in place and that prevent buckling of the vertical bars under compression load.

Concrete cover: The distance from the face of the concrete to the reinforcing steel; also referred to as fireproofing, clearance, or concrete protection.

Continuous beam: A beam that extends over three or more supports (including end supports).

Girder: The principal beam supporting other beams.

Reinforced concrete: Concrete that has been placed around some type of steel reinforcement material. After the concrete cures, the reinforcement provides greater tensile and shear strength for the concrete. Almost all concrete is reinforced in some manner.

Retaining wall: A wall that has been reinforced to hold or retain soil, water, grain, coal, or sand.

Simple beam: A beam supported at each end (two points) and not continuous.

Span: The horizontal distance between supports of a member such as a beam, girder, slab, or joist; also, the distance between the piers or abutments of a bridge.

Stirrups: Reinforcing bars used in beams for shear reinforcement; typically bent into a U shape or box shape and placed perpendicular to the longitudinal steel.

Concrete is arguably one of the most important construction materials. Properly installed and reinforced, it can safely serve as the supporting structures for large buildings, bridges, roads, and dams. Without proper reinforcement, however, concrete structures are accidents waiting to happen. Skilled and knowledgeable craftworkers are required to select, place, and tie steel reinforcing bar (rebar) and welded-wire fabric reinforcement in concrete formwork for foundations, walls, floors, beams, columns, and pilings.

Concrete has good compressive strength, but it is relatively weak in tension or if subjected to lateral or shear forces. Many kinds of proprietary reinforcement have been used for concrete in the past. Today, steel is generally the material used. This is because it has nearly the same temperature expansion and contraction rate as concrete. In addition, modern reinforcement conforms to American Society for Testing and Materials (ASTM) International standards that govern both its form and the types of steel used. As an alternative to steel reinforcement, fibers made from steel, fiberglass, or plastic such as nylon are sometimes added to concrete mixtures to provide reinforcement.

As discussed in the module *Properties of Concrete*, concrete is a mixture of cement, fine and coarse aggregates, water, and possibly one or more admixtures. By varying the proportions of the mixture, concrete with different compressive strengths can be obtained. These strengths typically vary from about 2,000 to 6,000 pounds per square inch (psi). Concrete is also available in higher strengths. Concrete usually sets firm in a matter of hours and typically attains design strength in about 28 days. Reinforced concrete is a combination of concrete and steel. Reinforced concrete combines the compression resistance of concrete with the tension and shear resistance of steel.

The adhesion of concrete to the surface of steel reinforcing bars (rebar) and/or welded-wire reinforcement and the resistance provided by the bar deformation, or lugs, keep the bars from slipping through the concrete. This adhesion is called the concrete bond; it makes the two materials act as one.

One of the primary purposes of reinforcing steel is to control cracking caused by tension and shear loads on the concrete. Cracking can also be caused by expansion and contraction of the concrete due to temperature changes and concrete shrinkage. Concrete has a high compressive strength but a low tensile strength, so shrinking of concrete as it cures, along with bending or shear forces, can cause cracking. To avoid cracking, concrete is reinforced with steel, which has a very high tensile strength, so that in combination the final product can resist forces from any direction. Although some cracking is inevitable, the use of reinforcing steel results in small cracks rather than large ones. Unreinforced concrete is likely to develop large cracks, which can lead to concrete failure.

In some instances, reinforcing steel is used strictly to control cracking in a slab, and does not provide any structural support. Welded-wire fabric reinforcement is often used for this purpose, but #3 rebar (in-lb) on 12" centers may be used instead (see the section *Reinforcing Bars* for information on rebar sizes). In that application, the rebar is referred to as temperature steel.

Only the correct amount and type of reinforcement, placed in the correct locations, can serve the intended purpose. Concrete reinforcement must always be selected and placed according to the engineer's drawings and specifications.

1.1.0 Resistance of Forces by Reinforcing Bars

Reinforcing bars are most effectively used in the following applications:

- Simple beams *(slabs, joists, and girders)* – *Figure 1* shows that the top half of the beam is in compression and the bottom half is in tension, so the steel is placed in the lower half, far enough from the bottom to achieve the proper amount of concrete cover (discussed in the section *Working with Reinforcing Steel*). To resist diagonal tension, stirrups are placed vertically across the beam. The stirrups are spaced more closely near the support and farther apart near the middle of the span to offset the force of shear, as shown in *Figure 2*.
- Continuous beams – These are beams that deflect downward between supports and have an upward thrust over the supports. Steel is re-

quired at the bottom between supports and at the top over supports, as shown in *Figure 3*.
- *Overhang, interior support, and cantilevered beams* – Tension bars must be placed in the top of the overhang and cantilever and carried back into the main span or support. As shown in *Figure 4*, a reinforcing steel framework, known as a beam cage, may be constructed to provide support for the concrete.

> **NOTE**
> The deflection shown in Figures 1, 3, and 4 has been exaggerated for illustration purposes.

- *Cantilevered* retaining walls – Main bars are required on the side toward the earth. See *Figure 5*.

Reinforcing Bars

Reinforcing steel is one means of providing concrete with shear and tension strength. Here, bundles of reinforcing bar are being readied for a concrete slab.

27304-14_SA01.EPS

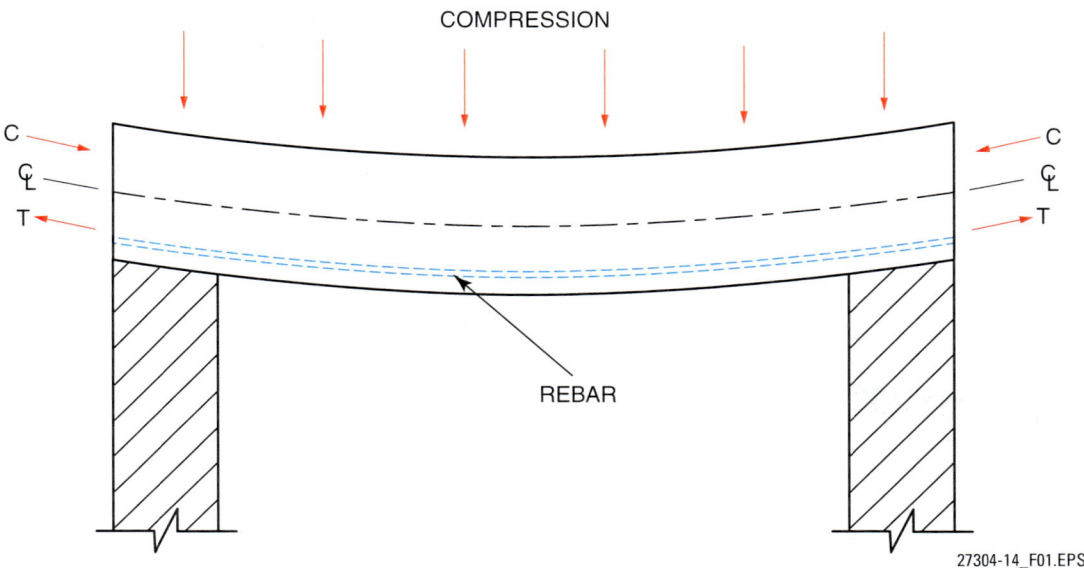

27304-14_F01.EPS

Figure 1 Rebar placement.

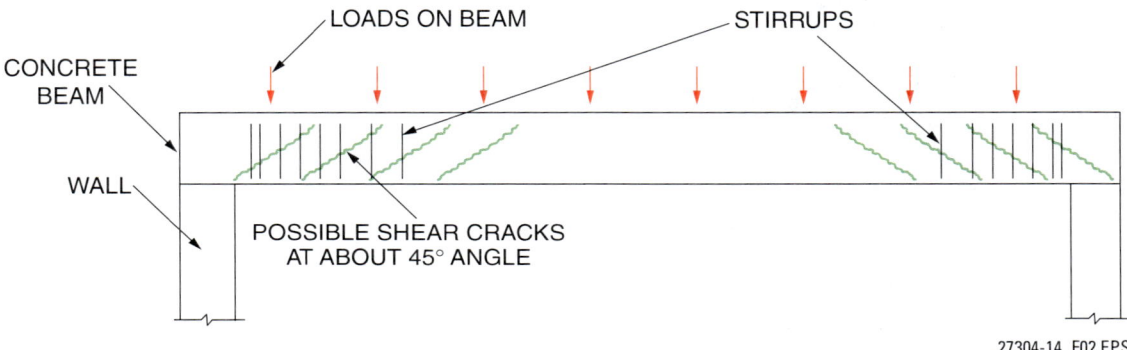

27304-14_F02.EPS

Figure 2 Stirrup placement.

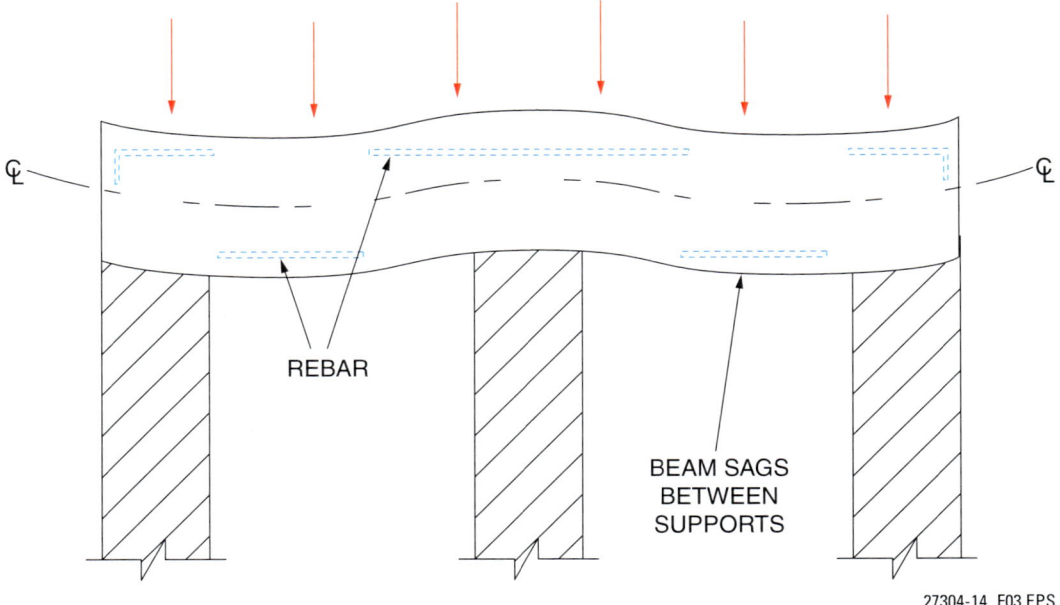

27304-14_F03.EPS

Figure 3 Continuous beam.

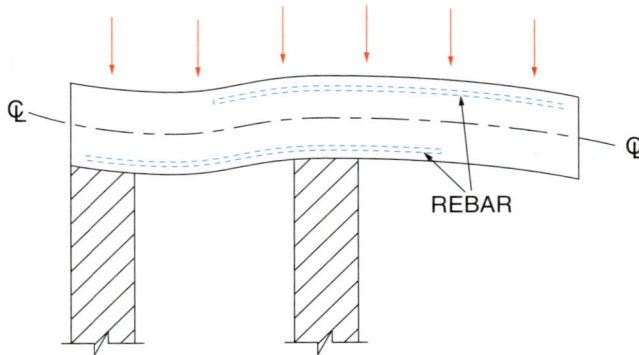

27304-14_F04.EPS

Figure 4 Cantilevered beam and beam cage.

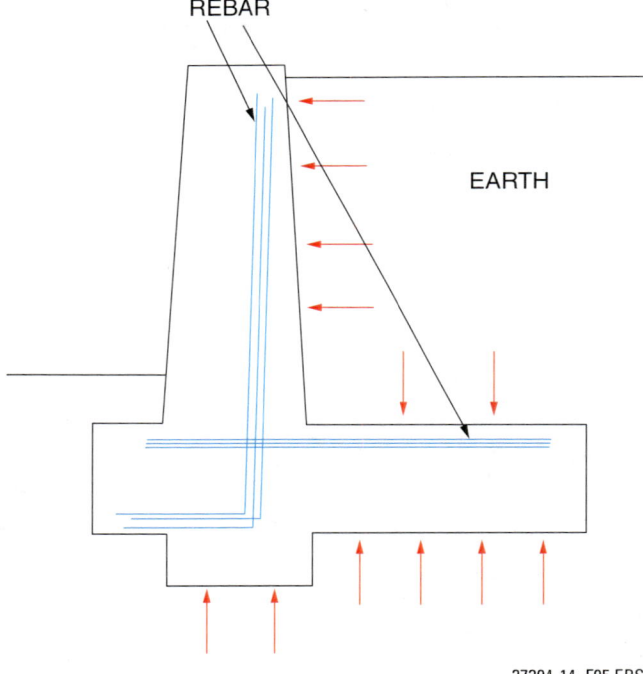

27304-14_F05.EPS

Figure 5 Cantilevered retaining wall.

- *Continuous footings* – These carry column loads at two or more points. Straight bars are placed near the top of the slab between the columns. Truss bars are placed under the column ends. Bottom crossbars prevent curling of the concrete.
- *Spread footings* – Bars are placed in two directions (at right angles to each other) and located a prescribed distance from the bottom of the footing, as shown in *Figure 6*. (Note that the deflection is exaggerated.)
- *Inside corners* – Reinforcing bars extend past the corner from each direction and are hooked for anchorage, if necessary.
- *Stairs and landings* – Bars continue across the tension point and are bent into the stair and landing slabs. See *Figure 7*.
- *Columns* – Reinforcing steel for compression forces is most commonly used in columns. If concrete alone were used, the column height would be very limited. Reinforcing bars are about 20 times stronger than an equivalent area of concrete, so they are used to carry part of the column load. Vertical column bars are in compression and will buckle if not restrained. Column ties (*Figure 8*) or column spirals act to prevent buckling.

1.2.0 Use of Reinforced Structural Concrete

Reinforced structural concrete has various applications in multistory building frames and floors, walls, shell roofs, folded plates, bridges, and prestressed or precast elements of all types. The architectural expression of form combined with functional design can be readily achieved with reinforced structural concrete. Architects, engineers, and contractors recognize that there are inherent economic and production values in the use of reinforced structural concrete, as evidenced by the many structures in which it is used.

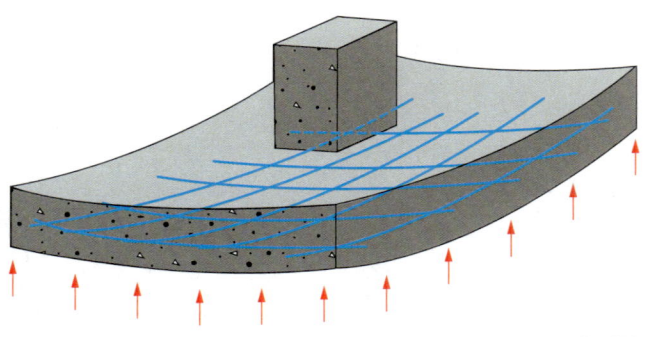

27304-14_F06.EPS

Figure 6 Spread footing.

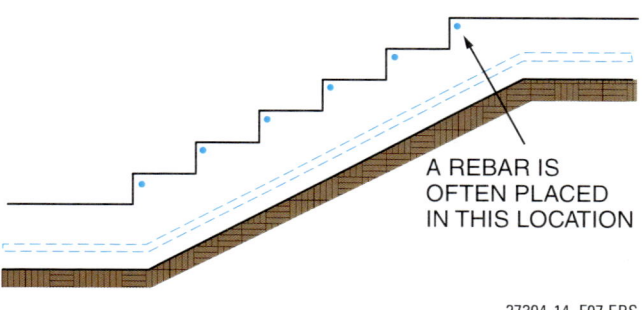

A REBAR IS OFTEN PLACED IN THIS LOCATION

27304-14_F07.EPS

Figure 7 Stairway.

1.2.1 Buildings

Building frames and floors of reinforced concrete are constructed using the joist floor system depicted in *Figure 9*. The illustration shows an exterior rectangular and an interior round column. The reinforced concrete joist floor shown in the figure consists of a series of ribs (small beams) with a top slab, which are all cast at one time with the supporting beams.

Special reinforced concrete construction includes arches, shells, and domes. A concrete arch is curved in one direction only, and the arch thickness is variable, being thinner at the center and thicker toward the supports. A barrel-shell roof is also curved in one direction, but the spans are supported between rigid frames. A dome roof is a slab of double curvature.

Curtain walls are usually prefabricated off site, then delivered for placement. Curtain walls

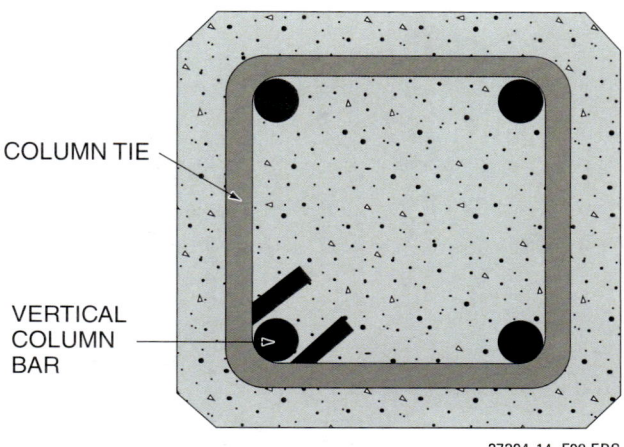

COLUMN TIE

VERTICAL COLUMN BAR

27304-14_F08.EPS

Figure 8 Column ties.

consist of one or more layers of vertical and horizontal reinforcement in a wall supported by the structural steel or concrete frame of the building, independent of the wall below (*Figure 10*).

1.2.2 Bridges

Some bridge construction units resemble those used in building construction and serve similar purposes. These units include footings, piers, caissons, walls, beams, and slabs.

A beam bridge (*Figure 11*) is commonly used for short spans. If a single span is used, the end supports are called abutments or end bents. When

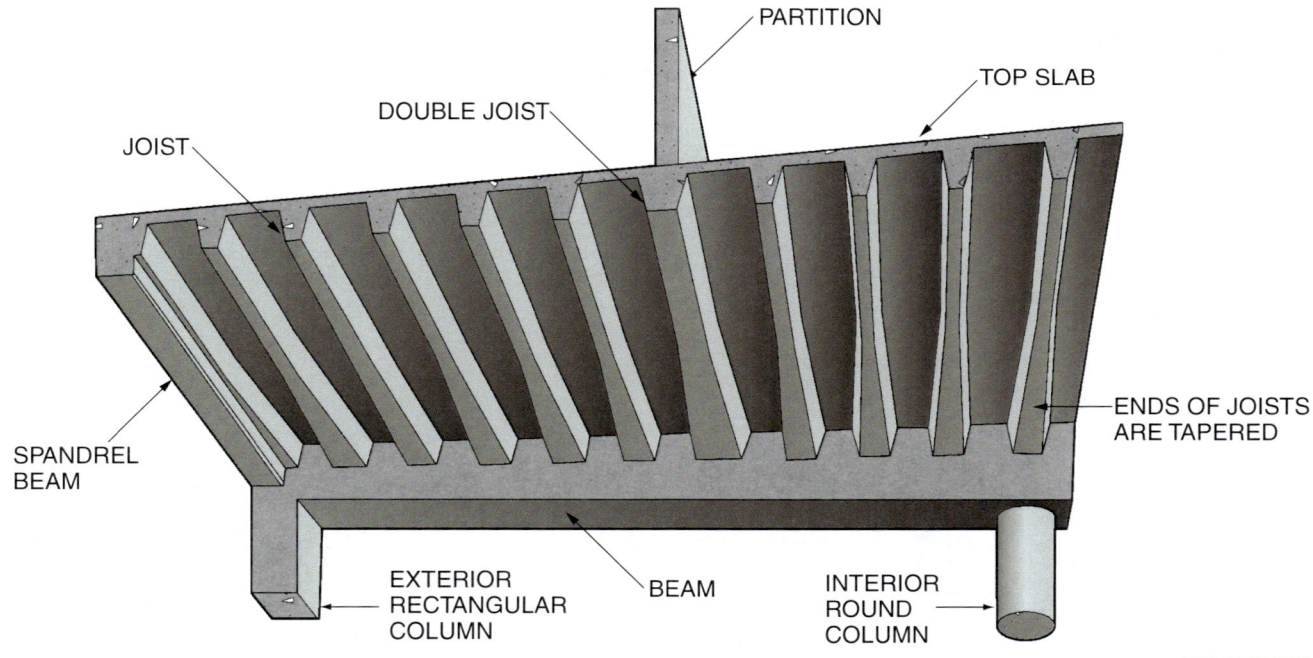

PARTITION

DOUBLE JOIST

JOIST

TOP SLAB

SPANDREL BEAM

ENDS OF JOISTS ARE TAPERED

EXTERIOR RECTANGULAR COLUMN

BEAM

INTERIOR ROUND COLUMN

27304-14_F09.EPS

Figure 9 Joist floor system.

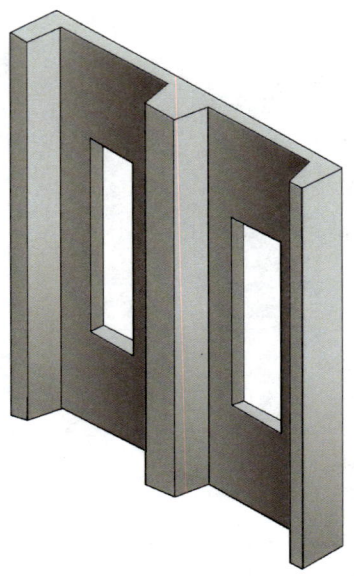

27304-14_F10.EPS

Figure 10 Precast wall unit.

two or more spans are used, the intermediate supports are called piers or intermediate bents.

A rigid-frame bridge (*Figure 12*) generally consists of footings, walls, and a deck slab cast as a unit, which forms a U-shaped element.

The arch bridge has many variations, one of which is the multiple-span, open-spandrel bridge configuration. The box-girder bridge (*Figure 13*) is

supported by abutments and piers in a manner similar to that of the beam bridge.

1.3.0 Posttensioned Concrete

When a load is applied to a conventional concrete slab such as an elevated parking garage deck, the concrete will tend to sag and may develop cracks. The use of rebar in the slab tends to combat this problem, but is not enough to prevent cracking under heavy load.

The solution is posttensioning, in which a steel tendon is placed in the form with the ends protruding from it. The tendon is placed in accordance with a posttensioning profile drawing developed by a qualified engineer. *Figure 14* is an example of posttensioning profiles for posttensioned beams. The multispan-beam profile diagram shows how the tendons go across the top of a column. A tendon consists of a bar or strand, along with its associated anchoring hardware and sheath. The strand is typically made from steel wires twisted together. Once the concrete has hardened around the tendon, one end of the tendon is anchored off, and the other end is tensioned using a special hydraulic jack (*Figure 15*). When the desired tension has been reached, the other anchor is secured.

There are two types of posttensioning: bonded and unbonded. In bonded posttensioning, a steel

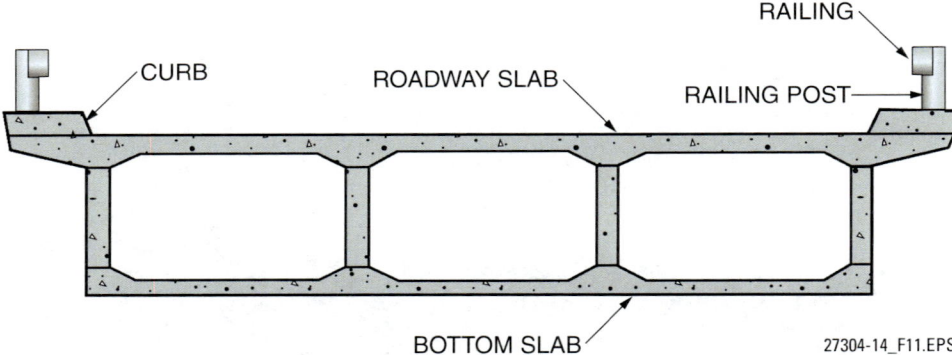

27304-14_F11.EPS

Figure 11 Beam bridge.

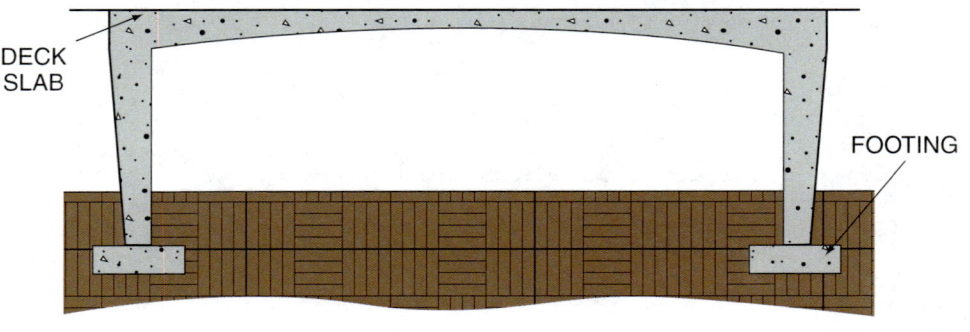

27304-14_F12.EPS

Figure 12 Rigid-frame bridge.

27304-14_F13.EPS

Figure 13 Box-girder bridge.

or plastic duct is inserted in the form. After the concrete is placed in the forms and has hardened, the strand is threaded through the duct. A tensioning force is applied, and then the duct is filled with grout.

In an unbonded system, the strand is covered with a corrosion-inhibiting grease and encased in a waterproof plastic sheath. The entire assembly is placed into the form before the concrete is placed. In some cases, the tendon sheaths are tied to the rebar for support while the concrete is placed (*Figure 16*).

The twisted-wire strand is used in large structures. The threaded bar is more common in smaller structures. A bearing plate and nut are used to anchor the bar.

27304-14_F15.EPS

Figure 15 Tendons being tensioned.

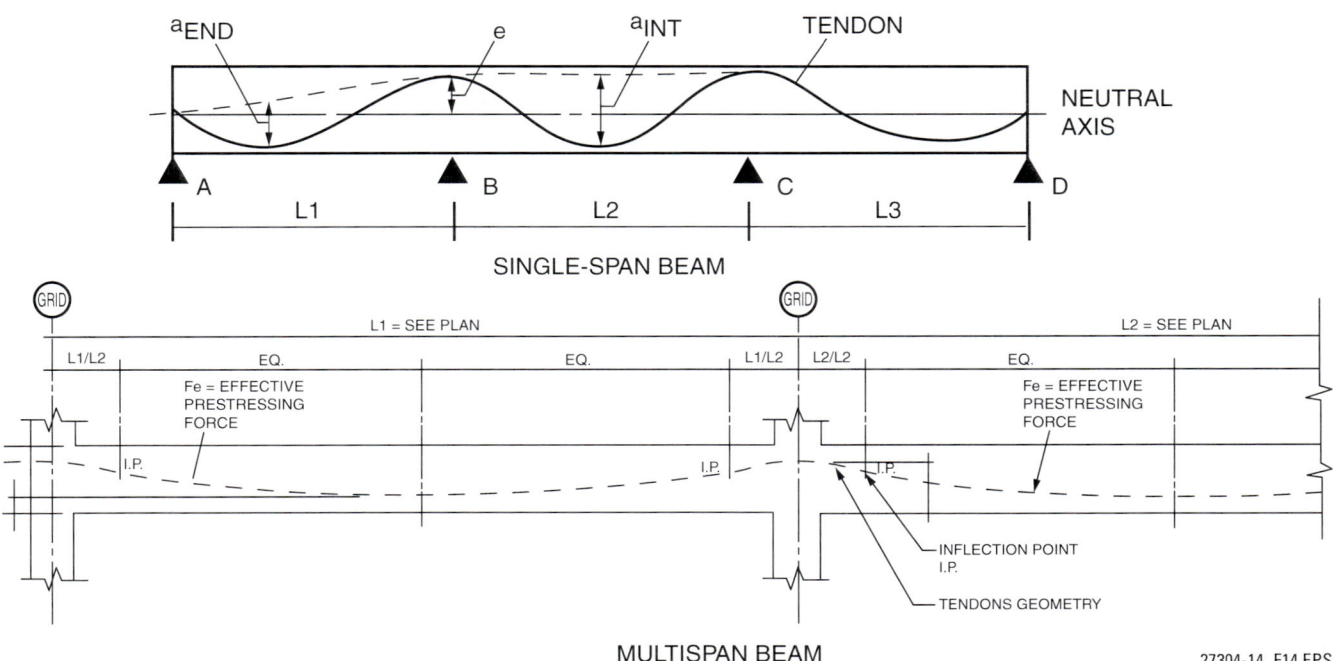

Figure 14 Examples of tendon profiles.

27304-14_F16.EPS

Figure 16 An unbonded posttensioned system.

2.0.0 WORKING WITH REINFORCING STEEL

Objective

Describe the general requirements for working with reinforcing steel, including tools, equipment, and fabricating methods.

   a. List general safety precautions when working with reinforcing steel.

   b. Describe the general characteristics of reinforcing steel.

   c. Discuss how reinforcing steel is fabricated.

   d. Explain the purpose of bar supports.

   e. Explain how welded-wire fabric reinforcement is used to reinforce concrete.

Trade Terms

Bar list: A bill of materials for a job site that shows all bar quantities, sizes, lengths, grades, placement areas, and bending dimensions to be used.

Bundle of bars: A bundle consisting of one size, length, or mark (bent) of bar, with the following exceptions: very small quantities may be bundled together for convenience, and groups of varying bar lengths or marks that will be placed adjacent to one another may be bundled together.

CRSI: Concrete Reinforcing Steel Institute.

Dowel: A bar connecting two separately cast sections of concrete. A bar extending from one concrete section into another is said to be doweled into the adjoining section.

Hickey bar: A hand tool with a side-opening jaw used in developing leverage for making in-place bends on bars or pipes.

Hook: A 180-degree (semicircular) or 90-degree turn at the free end of a bar to provide anchorage in concrete. For stirrups and column ties only, turns of either 90 degrees or 135 degrees are used.

Inserts: Devices that are positioned in concrete to receive a bolt or screw to support shelf angles, machinery, etc.

Pitch: The center-to-center spacing between the turns of a spiral.

Placing drawings: Detailed drawings that give the bar size, location, spacing, and all other information required to place the reinforcing steel.

Sleeve: A tube that encloses a bar, dowel, anchor bolt, or similar item.

Tie: A reinforcing bar bent into a box shape and used to hold longitudinal bars together in columns and beams. Also known as stirrup ties.

Tie wire: Wire (generally #16, #15, or #14 gauge) used to secure rebar intersections for the purpose of holding them in place until concreting is completed.

Craftworkers are employed by a contractor to place reinforcing bars in the concrete forms as indicated on the rebar placing drawings. One of the first jobs an apprentice will likely encounter will be the unloading, storing, and handling of reinforcing bars.

Bars must be placed carefully and accurately in order to conform to the requirements of the placing drawings. The engineer's structural drawings often have instructions for certain work to conform to a standard typical detail or note. Placing drawings must clearly indicate such items as top and bottom bars, bars hooked around other bars, and on which side or face of the member the bar is to be located. Bars must be fitted around sleeves, inserts, holes, and other openings as shown on the plans. The craftworker, supervisor, contractor, and job inspector all have the responsibility to see that bars are correctly placed.

So that the strength of any concrete member is not adversely affected, bars must be placed and held in position as shown on the placing drawings. If, for example, the top bars are lowered or the bottom bars are raised $\frac{1}{2}$" more than specified in a 6"-deep slab, the load-carrying capacity could be reduced by 20 percent. A sufficient number of supports should be used in order to place the bars in their specified locations.

Considerable care is necessary if top bars are to be held in the proper position. These bars often interfere with other bars at right angles to the top bars, or with other items (often called embedments) buried in the slab, such as ducts and wiring conduits. The designer usually takes care of these potential problems, but sometimes the top bars cannot be placed to the $\frac{1}{2}$" tolerance and must be relocated.

Bars may receive heavy abuse on the job. In order to ensure that they stay in position, they should be securely wired and held against movement. The supervisor, architect, or other authority inspects and checks the steel for proper placement before the concrete is placed. In all cases, no variation may be made from the placing drawings without proper authorization.

The structural engineer determines the amount of concrete cover for each part of the job. This decision is based on various considerations, such as the building codes, fire hazards, exposure to weather, and the possibility of corrosion. Where not specified, however, the minimum standards established by the American Concrete Institute (ACI) should be used as guidelines.

ACI 318-95 (Building Code Requirements for Structural Concrete) section 7.7.1 includes the following guidelines for concrete cover:

- 3" at sides where concrete is cast against earth and on bottoms of footings or other principal structural members where concrete is deposited on the ground
- 2" for bars larger than #5 where concrete surfaces would be exposed to the weather after removal of forms or would be in contact with the ground; 1½" for #5 bars and smaller
- 1½" over spirals and ties in columns
- 1½" to the nearest bars on the top, bottom, and sides of beams and girders
- ¾" for #11 and smaller bars on the top, bottom, and sides of joists, and on the top and bottom of slabs where concrete surfaces are not directly exposed to the ground or weather; 1½" for #14 and #18 bars
- ¾" from the faces of all walls not directly exposed to the ground or weather for #11 and smaller bars; 1½" for #14 and #18 bars

Always refer to the structural notes for clearance and coverage requirements.

2.1.0 General Safety Precautions

Developing an overall safety consciousness is very important in any phase of construction work. The following general safety precautions apply to the placing and tying of reinforcing bars:

- Wear proper PPE, including hard hats, boots, and leather gloves.
- Do not wear loose or ragged clothing.
- Ensure your footing is always solid.
- Wear approved fall protection equipment.
- Keep the work area clean.
- Never drop any material to a lower level.
- Block piles of reinforcing steel to prevent sideways movement.
- Wear goggles with the proper shade of lens when cutting with a torch.
- Know and observe the proper hand signals for hoisting equipment. Only qualified riggers and signal persons can perform these tasks.
- Never ride on any material being hoisted.
- Lift bars or bundles properly to avoid strain.
- Always know the location of co-workers.

- Pull any projecting nails.
- When carrying bars with a partner, make sure the load is balanced. Each person should be positioned about one-quarter the length of the rebar from each end. Lift and set all bars in unison.
- Never land hoisted bundles or drop carried bars on formwork.
- Be alert for concrete buggies or any hoisted material that is swinging.
- Do not hoist bundles by the #9-gauge wire wrappings that tie bars together. Use proper slings and chokers. Double chain slings are recommended; nylon slings are not allowed for this application.
- Adequately guy and support reinforcing steel for vertical structures such as piers and columns to prevent collapse.
- Brake and secure rolled-out wire reinforcement to prevent dangerous recoiling action.
- Report all unsafe conditions.
- Always use American National Standards Institute (ANSI)-approved rebar caps on exposed ends of bars.
- Bend down loose ends of tie wires with pliers after each tie. If this is not done, the wire may puncture gloves, boots, or skin.
- Wear gloves when handling rebar. Make sure the gloves are heat resistant; rebar can get very hot lying in the sun.

2.2.0 Reinforcing Bars

Reinforcing bars, often called rebar, are available in several grades. These grades vary in yield strength, ultimate strength, percentage of elongation, bend-test requirements, and chemical composition. In addition, reinforcing bars can be coated with different compounds, such as epoxy, for use in concrete where corrosion could be a problem. To obtain uniformity throughout the United States, ASTM International has established standard specifications for these bars. These grades will appear on bar-bundle tags, in color coding, in rolled-on markings on the bars, and/or on bills of materials. The specifications are as follows:

- *ASTM A615, Standard Specification for Deformed and Plain Carbon-Steel Bars for Concrete Reinforcement*
- *ASTM A996, Standard Specification for Rail-Steel and Axle-Steel Deformed Bars for Concrete Reinforcement (this standard replaces A616 and A617)*

- *ASTM A706, Standard Specification for Low-Alloy Steel Deformed Bars and Plain Bars for Concrete Reinforcement*

The standard configuration for reinforcing bars is the deformed bar. Different patterns may be impressed on the bars, depending on which mill manufactured them, but all are rolled to conform to ASTM specifications. The deformation improves the bond between the concrete and the bar, and prevents the bar from moving in the concrete.

Plain bars are smooth and round without deformations on them and are used for special purposes, such as for dowels at expansion joints where the bars must slide in a sleeve, and for expansion and contraction joints in highway pavement.

Deformed bars are designated by a number in 11 standard sizes (metric or inch-pound), as shown in *Table 1*. The number denotes the approximate diameter of the bar in eighths of an inch or in millimeters. For example, a #5 bar has an approximate diameter of ⅝". The nominal dimension of a deformed bar (nominal does not include the deformation) is equivalent to that of a plain bar having the same weight per foot.

As shown in *Figure 17*, bar identification is accomplished by ASTM specifications, which require that each bar manufacturer roll the following information onto the bar:

- Letter or symbol to indicate the manufacturer's mill
- Number corresponding to the size number of the bar (*Table 1*)
- Symbol or marking to indicate the type of steel (*Table 2*)
- Marking to designate the grade

The grade represents the minimum yield (tension strength) measured in kilopounds per square inch (ksi) that the type of steel used will withstand before it permanently stretches (elongates) and will not return to its original length (*Table 3*). Today, Grade 420 is the most commonly used rebar. Bars are normally supplied from the mill bundled in 60' lengths.

Bar fabrication is accomplished for straight bars by cutting them to specified lengths from the 60' stock. Bent bars are cut to length the same as straight bars, and then they are assigned to a bending machine that is best suited for the type of bend and size of the bar. *Table 1* provides size and weight information for various bars so that proper handling and bending equipment can be selected.

Uncoated reinforcing steel that has not been contaminated by oil, grease, or preservatives will normally rust when stored, even for short lengths of time under cover. A number of studies, some conducted over 70 years ago, have shown that rust and tight mill scale actually improve the bond between the steel and the concrete. Other studies have shown that normal handling (moving, bending, etc.) of extremely rusted reinforcement steel prepares it sufficiently for proper bonding with concrete without additional effort to remove the rust.

To reduce congestion in cast-in-place construction when multiple hooked-end bars converge from several directions, *ASTM A970* permits the use of headed reinforcing bars (HRBs). The head is a rectangular or round steel plate and can be welded, threaded, or forged on the end of the rebar.

Table 1 ASTM Standard Metric and Inch-Pound Reinforcing Bars

| Bar Size | | Nominal Characteristics* | | | | | |
| | | Diameter | | Cross-Sectional Area | | Weight | |
Metric	[in-lb]	mm	[in]	mm	[in]	kg/m	[lbs/ft]
#10	[#3]	9.5	[0.375]	71	[0.11]	0.560	[0.376]
#13	[#4]	12.7	[0.500]	129	[0.20]	0.944	[0.668]
#16	[#5]	15.9	[0.625]	199	[0.31]	1.552	[1.043]
#19	[#6]	19.1	[0.750]	284	[0.44]	2.235	[1.502]
#22	[#7]	22.2	[0.875]	387	[0.60]	3.042	[2.044]
#25	[#8]	25.4	[1.000]	510	[0.79]	3.973	[2.670]
#29	[#9]	28.7	[1.128]	645	[1.00]	5.060	[3.400]
#32	[#10]	32.3	[1.270]	819	[1.27]	6.404	[4.303]
#36	[#11]	35.8	[1.410]	1006	[1.56]	7.907	[5.313]
#43	[#14]	43.0	[1.693]	1452	[2.25]	11.38	[7.65]
#57	[#18]	57.3	[2.257]	2581	[4.00]	20.24	[13.60]

*The equivalent nominal characteristics of inch-pound bars are the values enclosed within the brackets.

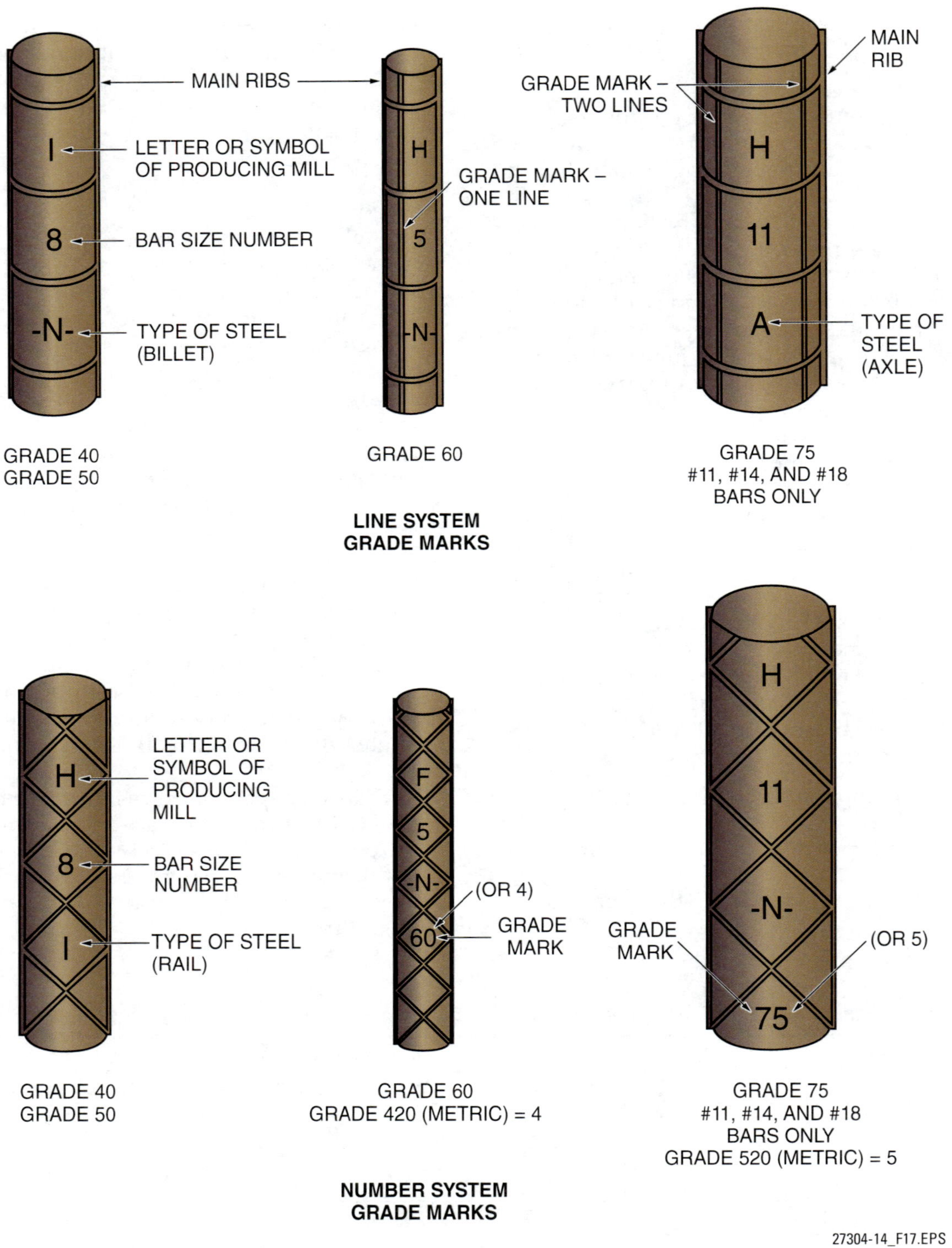

LINE SYSTEM
GRADE MARKS

NUMBER SYSTEM
GRADE MARKS

27304-14_F17.EPS

Figure 17 Reinforcing bar identification.

ASTM A955M provides information on stainless steel reinforcing bars (SSRBs). These bars, which have been used for some time, are used in highly corrosive environments or when nonmagnetic reinforcement must be used. Specification *A955M* defines these rebars in metric units, and the requirements parallel those in *ASTM A615*.

2.3.0 Fabrication

In a discussion of reinforcing steel, the terms *fabricating* and *manufacturing* should not be confused. The manufacturing of reinforcing steel produces a deformed bar from a specific type and grade of steel that is rolled to a stock straight length. The fabrication of reinforcing steel is the process of

NCCER – *Carpentry Level Three* 27304-14

Table 2 Reinforcing-Bar Steel Types

Symbol/Marking	Type of Steel
A	Axle (ASTM A617)
S or N	Billet (ASTM A615)
I or IR	Rail (ASTM A616)
W	Low-alloy (ASTM 706) (for welded lap, butt joints, etc.)

Table 3 Reinforcing-Bar Grades

Grade	Identification	Minimum Yield Strength
40 and 50	None	40,000 to 50,000 psi (40 to 50 ksi)
60	One line or the number 60	60,000 psi (60 ksi)
70	Two lines or the number 70	70,000 psi (70 ksi)
420	The number 4	60,000 psi (60 ksi)
520	The number 5	75,000 psi (75 ksi)

cutting and bending reinforcing bars to suit the particular needs of the job as required by the placing drawings. Most fabrication of reinforcing bars for commercial structures is done in shops, but a certain amount of it is done in the field.

2.3.1 Tools

The tools and equipment used by craftworkers working with reinforcing steel include the following:

- Hard hat
- ANSI-approved footwear
- 2" leather belt
- Tie-wire reel (needs to be good quality; don't skimp on the cost)
- Tool pouch
- Side-cutting pliers (wire cutters)
- Tape measure
- Keel holder (for holding soapstone, chalk, or crayon to mark bars)
- Leather-palm gloves (heat and puncture resistant)
- Safety glasses
- Level
- Plumb bob

- Bolt cutters (for cutting rebar)
- Sledgehammer (for aligning rebar)
- Hickey bar for bending rebar (#5 and less)

For the job of cutting welded-wire fabric reinforcement or small-size wires, side-cutting pliers or a bolt cutter are used. Rods up to ⅝" in diameter are cut with some type of hand-operated machine or cutting torch. Larger rebars can be cut with a portable band saw, chop saw, or cutting torch.

2.3.2 Fabricated Bars

The American Concrete Institute (ACI) and the Concrete Reinforcing Steel Institute (CRSI) have standardized the most common types of bar bends and assigned each a number or a letter preceding a number. These designations are used by engineers, fabricators, inspectors, and ironworkers. Information concerning bar bends is usually found on placing drawings and bar lists.

Figure 18 illustrates the typical bar bends standardized by ACI. The circled letter and/or number adjacent to each bar-bend detail indicates its type.

Unless otherwise noted, all hooks are formed in accordance with the recommended sizes for 180-degree hooks, as specified in the latest edition of *ACI 315, Details and Detailing of Concrete Reinforcement*. The dimensions and geometry of standard hooks are shown in *Figure 19* and *Table 4*.

It is important not to omit any of the letters from a standard-bend type. Doing so will result in a different shape of bar. For example, a Type 4 bar without the A or G dimension would be a straight-truss bar.

Each dimension of the standardized bar bends has been assigned a letter. These dimensions are read out-to-out, that is, from the outside diameter of the bars.

2.3.3 Bar Lists

As indicated by its name, a bar list is a list of reinforcing bars contained in a shipment. It is prepared by the fabricator and will include a comprehensive list of reinforcing bars for an entire shipment or a separate list of bars for each truckload within a shipment. The fabricators receive their information from the construction drawings.

MMFX₂® Rebar

MMFX₂® rebar is a corrosion-resistant rebar that provides five times more corrosion resistance and twice the strength of standard rebar. While stainless steel rebar provides the corrosion resistance, it is also very expensive when compared to MMFX₂® rebar. MMFX₂® rebar is used in a variety of commercial, industrial, waterway, and transportation projects.

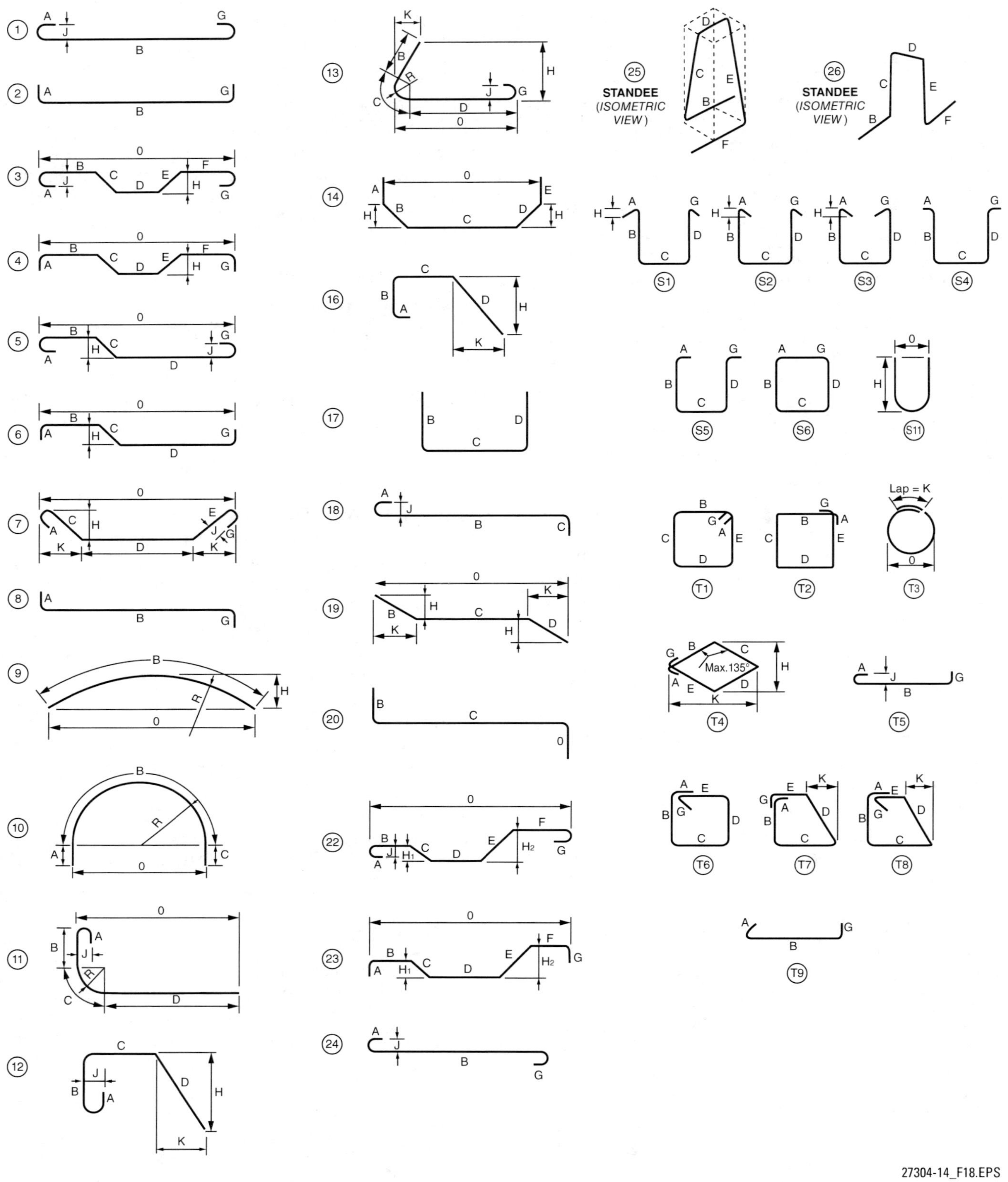

Figure 18 Typical bar bends.

RECOMMENDED END HOOKS, ALL GRADES OF STEEL

BAR SIZE	D		180° HOOKS		90° HOOKS
			A OR G	J	A OR G
#10 (#3)	60	(2 1/4")	125 (5")	80 (3")	150 (6")
#13 (#4)	80	(3")	150 (6")	105 (4")	200 (8")
#16 (#5)	95	(3 3/4")	175 (7")	130 (5")	250 (10")
#19 (#6)	115	(4 1/2")	200 (8")	155 (6")	300 (12")
#22 (#7)	135	(5 1/4")	250 (10")	180 (7")	375 (14")
#25 (#8)	155	(6")	275 (11")	205 (8")	425 (16")
#29 (#9)	240	(9 1/2")	375 (15")	300 (11 3/4")	475 (19")
#32 (#10)	275	(10 3/4")	425 (17")	335 (13 1/4")	550 (22")
#36 (#11)	305	(12")	475 (19")	375 (14 3/4")	600 (24")
#43 (#14)	485	(18 1/4")	675 (27")	550 (21 3/4")	775 (31")
#57 (#18)	610	(24")	925 (36")	725 (28 1/2")	1050 (41")

NOTE: All metric dimensions are in millimeters (mm). Numbers shown in parentheses indicate inch-pound sizes and dimensions.

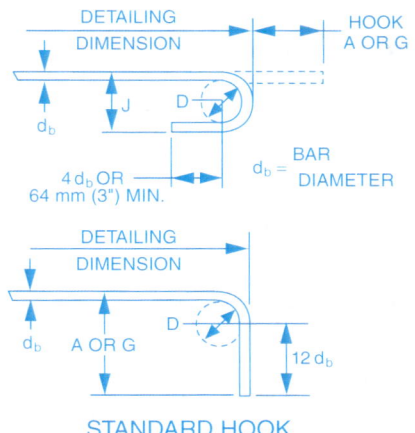

STANDARD HOOK

STIRRUP HOOKS (TIE BENDS SIMILAR)

BAR SIZE	D		90°	135°	
			A OR G	A OR G	H*
#10 (#3)	40	(1 1/2")	105 (4")	105 (4")	65 (2 1/2")
#13 (#4)	50	(2")	115 (4 1/2")	115 (4 1/2")	80 (3")
#16 (#5)	65	(2 1/2")	155 (6")	140 (5 1/2")	95 (3 3/4")
#19 (#6)	115	(4 1/2")	305 (12")	205 (7 3/4")	115 (4 1/2")
#22 (#7)	135	(5 1/4")	355 (14")	230 (9")	135 (5 1/4")
#25 (#8)	155	(6")	410 (16")	270 (10 1/4")	155 (6")

NOTE: All metric dimensions are in millimeters (mm). Numbers shown in parentheses indicate inch-pound sizes and dimensions. *H dimension is approximate.

SEISMIC STIRRUP/TIE

BAR SIZE	D		135° SEISMIC HOOK	
			A OR G	H*
#10 (#3)	40	(1")	110 (4 1/4")	80 (3")
#13 (#4)	50	(2")	115 (4 1/2")	80 (3")
#16 (#5)	65	(2 1/2")	140 (5 1/2")	95 (3 3/4")
#19 (#6)	115	(4 1/2")	205 (7 3/4")	115 (4 1/2")
#22 (#7)	135	(5 1/4")	230 (9")	135 (5 1/4")
#25 (#8)	155	(6")	270 (10 1/4")	155 (6")

NOTE: All metric dimensions are in millimeters (mm). Numbers shown in parentheses indicate inch-pound sizes and dimensions. *H dimension is approximate.

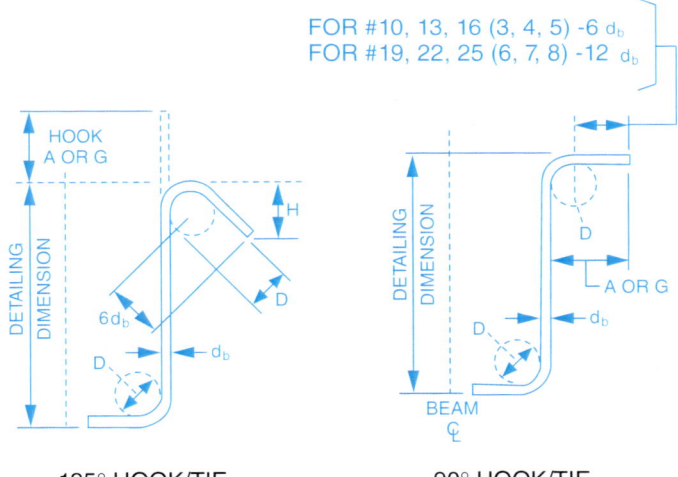

135° HOOK/TIE   90° HOOK/TIE

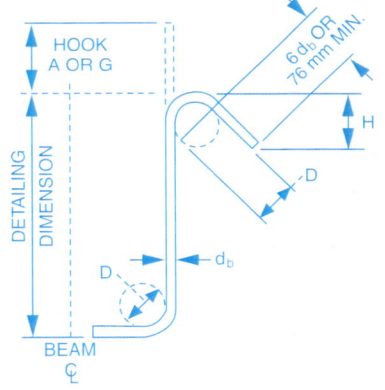

135° SEISMIC HOOK/TIE

27304-14_F19.EPS

Figure 19 Standard hook details for reinforcing bars.

A bar list is useful to everyone involved in placing and fabricating reinforcing steel. The fabricator uses it for bending, tagging, shipping, and invoicing. The field crew uses it for checking quantities when unloading a shipment, sorting bars, delivering the correct bars to the correct area of the job, and for the actual placing of the bars in the forms.

2.3.4 Bar-List Information

Fabricators generally use their own forms for bar lists. Sometimes straight bars and bent bars will be placed on separate sheets, or they may be placed on the same sheet. However, certain information will be common to all bar lists. The bars are first classified as one of four types, three of which refer to bent bars. These types are as follows:

Table 4 Standard Hook Specifications

| Material (ASTM Specification Number) | Bend Diameters | | | Bar Size | Grades |
| | ACI | | ASTM[3] | | |
	Standard[1]	Stirrup/Tie[2]			
A615/615M	6d[4]	4d	3.5d	10, 13, 16 (3, 4, 5)	300, 420 (40, 60)
	6d	6d	5d	19 (6)	300, 420, 520 (40, 60, 75)
	6d	6d	5d	22, 25 (7, 8)	420, 520 (60, 75)
	8.5d	N/A	7d	29, 32, 36 (9, 10, 11)	420, 520 (60, 75)
	10.7d	N/A	9d(90°)	43, 57 (14, 18)	420, 520 (60, 75)
A706/A706M	6d	4d	3d	10, 13, 16 (3, 4, 5)	420 (60)
	6d	6d	4d	19, 22, 25 (6, 7, 8)	420 (60)
	8.5d	N/A	6d	29, 32, 36 (9, 10, 11)	420 (60)
	10.7d	N/A	8d	43, 57 (14, 18)	420 (60)

[1]ACI standard bends 90° and 180°. Finished bend diameter affected by spring-back for bars 29 to 57 (9 to 18).
[2]ACI stirrup/tie bends 90° and 135° for bars 10 to 25 (3 to 8) only.
[3]ASTM bend tests 180° unless otherwise specified.
[4]d = the nominal diameter of the bar.

- *Straight bars* – This category consists of standard straight reinforcing bars.
- *Heavy bending bars* – This category consists of metric bar sizes #13 through #57 (in-lb sizes #4 through #18) that are bent at no more than six points in one plane. Single-radius bending is also grouped in this category.

> **NOTE**
> In the following discussion and throughout this module, metric bar sizes are listed first and in-lb bar sizes are noted in parentheses.

- *Light bending bars* – This category includes all #10 (#3) bars, all stirrups and ties, and all bars #13 through #57 (#4 through #18) bent at more than six points in one plane. It also includes all single-plane radius bending with more than one radius in any bar, or a combination of radius or other type of bending in one plane.
- *Special bending bars* – This category includes all bending to special tolerances, all radius bending in more than one plane, all multiple-plane bending with one or more radius bends, and all bending for precast units.

The bars are then grouped by sizes and lengths under each classification. The largest size and the longest length of each size will be listed first. Each bar list usually contains the following information:

- Name of project
- Customer name
- Job location
- Part of job
- Placing-drawing reference number
- Grades of steel
- Order number

2.3.5 Sample Bar List

Figure 20 illustrates a sample bar list that might be prepared by a fabricator and sent with the shipment of reinforcing bars to the job site. Notice that the top of the sheet contains all the necessary information about the project, customer, etc. Also, notice the divisions labeled Straight Bars, Light Bending, and Heavy Bending.

Each bar is described by the information found under the column headings. Most of the headings are self-explanatory, listing the quantity, bar size, and length of the bars. The mark refers to the location of the bent bars as found on the placing drawing. The bend type refers to one of the standard circled letter and/or number bar bends shown in *Figure 18*. Placement and dimensions of the bends are identified in columns A through H, J, and O as required, and as shown in *Figure 18*.

Bar lists can also include a safety data sheet that lists hazardous ingredients, physical data, fire and explosion information, and reactivity information for the metals used in the rebar.

2.3.6 Special Details of Fabrication

Special details of fabrication include the following:

- *Hooks* – Hooks refer to those parts of a bar with dimensions designated A or G by ACI standards. (Refer to *Figure 18* for examples of these designations.) All other dimensions are properly called bends. There are three types of hooks in general use, each described by the size of the angle the hook encompasses—a 90-degree hook, a 135-degree hook, and a 180-degree hook (refer to *Figure 19*).

- *Spirals* – Spirals are made of smooth bar stock or of wire shaped like a coil spring. They have spacers attached on opposite sides that keep the turns of the spiral at equal distances. Center-to-center spacing of the turns of a spiral is called pitch. Spirals are usually shipped collapsed and must be opened or expanded at the job site prior to assembly. The spacers should be arranged to provide equal pitch around the spiral column.
- *Radial bending* – Curved reinforcing bars are used for tanks, bins, culverts, domes, tunnels, and other curved structures. These bars are either fabricated in a shop or sprung into shape in the field, depending on the maximum radius of the bend. *Table 5* shows the maximum radii that are generally shop fabricated. Any bars of a larger radius will usually be shipped straight.

WARNING! Place rebar according to the structural drawings and specifications. Do not place according to field decisions or architectural drawings.

2.3.7 Tolerances in Fabrication

Tolerances are necessary in normal fabricating operations. The engineers typically allow for tolerances in their designs. The usual tolerances are as follows:

- *Straight bars* – ±1" (25.4 mm) in length.
- *Hooked bars* – #7 or smaller, the overall length may deviate by ±½" (12.7 mm); #8 or larger, the tolerance is ±1" (25.4 mm).
- *Truss bars* – #7 and smaller, the overall length may deviate by ±½" (12.7 mm); #8 bars and larger, the tolerance is ±1" (25.4 mm); for all bars, the H dimension may vary only 0" to ½" (0 mm to 12.7 mm).
- *Spirals* – ±½" (12.7 mm) tolerance.
- *Column ties* – ±½" (12.7 mm) tolerance.

Even though fabricators usually supply the job site with the proper number and types of bars, some field fabrication is generally necessary. This fabrication may range from straightening a bent bar or dowel to replacing a lost or badly damaged truss bar.

2.3.8 Bundling and Tagging

Straight bars are bundled and tagged. Other bars are bent first, then bundled and tagged. Unassembled spirals are bundled immediately after fabrication and tagged. Assembled spirals are tagged individually and shipped collapsed, if so designed.

Each bundle of bars should contain bars of one size, length, and mark. *Mark* is the term used to designate the part of the structure for which the bars are intended. Bundles are generally secured by wraps of #9-gauge wire spaced 10' to 15' apart, with a minimum of two ties per bundle. Each bundle should be tagged. The tag should be made of a durable material such as metal or rope fiber. Metal tags should be embossed; fiber tags should be marked with waterproof ink. In general practice, the tag will show the purchaser by name, address, and order number. The bundle of bars should be identified by the number of pieces, size and length of straight bars, mark number for bent bars, and grade of steel for both bent and straight bars. Tags for bent bars generally give information on bending dimensions.

Concrete Bridge

This cast-in-place, 1,345' concrete frame is made up of eight spans, including six interior spans of 177' each. It is supported by 7' octagonal columns set atop 8'-diameter piles.

27304-14_SA02.EPS

				ORDER NO.		JC1147									
PROJECT...................			HOCKEY ARENA			ORG. NO.		WQ31							
CUSTOMER................			M & G CONSTRUCTION			SHEET		1 OF 3							
LOCATION..................			MIAMI, FLORIDA			DATE		5/10/01							
MAT'L FOR..................			1ST FLOOR BEAMS & COLUMNS			MADE BY..............JAC		CHECKED BY.....ABC							

ITEM	QTY.	BAR SIZE	FINAL LENGTH	MARK	BEND TYPE	A	B	C	D	E	F	G	H	J	O
1			STRAIGHT BARS												
2	4	#7	22-0												
3	4		17-6												
4															
5	2	#5	28-3												
6	2		17-6												
7			HEAVY BENDING												
8	2	#9	36-0	1B 901	3		10-0	2-3	12-4	2-3	9-2		1-7		
9	2		35-7	1B 902	3		10-0	2-3	12-4	2-3	9-2		1-7		
10															
11	2	#8	23-8	1B 801	1	1-1	22-7								
12															
13	2	#7	25-2	1B 703	3	10	2-3	2-81/2	9-10	2-81/2	6-10		1-11		
14															
15	2	#6	26-2	1B 601	3	8	5-7	2-7	8-6	2-7	5-7	8	1-10		23-4
16															
17			LIGHT BENDING												
18	22	#4	5-6	U401	S2	4 1/2	1-11	11	1-11			4 1/2			
19	34		5-2	U402	S1	4 1/2	1-9	11	1-9			4 1/2			
20															
21	26	#3	6-4	U301	S2	4	2-6	8	2-6			4			
22	24		6-0	U302	T1	4	2-0	8	2-0	8		4			
23															
24	12	#2	6-11	U201	T1	3 1/2	1-4	1-5	1-9	1-5		3 1/2			
25	20		3-11	U202	T1	3 1/2	10	10	10	10		3 1/2			

27304-14_F20.EPS

Figure 20 Sample bar list.

Table 5 Maximum Prefabricated Radii

Bar Size	Maximum Prefabricated Radius
#3	10'
#4	10'
#5	15'
#6	40'
#7	40'
#8	60'
#9	90'
#10	110'
#11	110'
#14	180'
#18	300'

WARNING!

Do not rig to the #9-gauge wire bundle wrap; rig to the rebar instead.

Many rebar fabricators use computer-based applications that automatically generate bar lists and print labels for rebar bundles. *Figure 21* shows another example of a bar list, and *Figure 22* shows two labels made from the bar list. Note that the labels are directly related to two of the entries on the bar list.

WARNING!

OSHA standard 29 *CFR* 1926.701(b) requires that all protruding reinforcing steel, onto and into which workers could fall, shall be guarded to eliminate the hazard of impalement.

27304-14_SA03.EPS

2.4.0 Bar Supports

Bar supports, sometimes called accessories, are used to support, hold, and space reinforcing bars and mats or wire reinforcement before and during concrete placement. Bar supports are made from steel, concrete, or plastic. When used with coated reinforcement steel, the supports should be coated with the same material or made of concrete or plastic to prevent corrosion. A sufficient number of supports and the correct size of the supports must be used to prevent the reinforcement from shifting out of position or deforming when the concrete is placed.

NOTE

For the rebar to function as designed, bar-support sizing is critical for bar placement and positioning as intended by the engineer.

2.4.1 Steel-Wire Bar Supports

Steel-wire bar supports are divided into five classes based on how well the support will prevent rust spots or similar blemishes from forming on the surface of the concrete. These classes are as follows:

- *Class A, bright basic* – Class A offers no protection against rusting. Therefore, it is used in situations where surface blemishes can be tolerated.
- *Class B, pregalvanized* – Class B offers minimal protection against rusting. It is used where nominal protection for a short amount of time is required.
- *Class C, plastic protected* – Class C offers moderate protection against rusting. It is used in situations where the surface of the concrete will be subjected to moderate exposure or where sandblasting or light grinding of the concrete surface is required.
- *Class D, stainless protected* – Class D offers more protection against rusting than Class C and is used in the same situations.
- *Class E, special stainless* – Class E is used where the concrete surface will be exposed to moderately severe conditions or where heavy grinding or sandblasting of the concrete surface is required.

Steel Reinforcement Protection

Steel reinforcement of any kind must be covered by enough concrete to be adequately protected; otherwise, the steel will rust, causing damage to the concrete. Some examples of minimum concrete coverage are:

- *Footings* – 3"
- *Concrete surface exposed to weather* – 2" for bars larger than #5, 1½" for bars #5 and smaller
- *Slabs, walls, joists* – ¾"
- *Beams and girders* – 1½"

Client: SAMPLE CLIENT 1   Project: SAMPLE IMPERIAL PROJECT
Address: New York, Elm Street, 67   Type:
Code: 2001-1   Number: 1   Module: FACTORY BLDG.
Description: FIRST FLOOR FRAMING PLAN

Mark	Total	Size	Long	Wt. lb	Wt. lb/Page = 951,05
					FIRST FLOOR
5K9	4	#5	6'4½"	26,60	0..> 6'4½"
4K10	5	#5	7'8"	39,98	2..> 10" 6' 10"
4K12	3	#9	12'4½"	126,01	2..> 1'-7" 9'2½"
4K13	3	#9	6'10"	69,70	2A..> 1'-7" 5'-3"
3K8	2	#3	10'10"	8,15	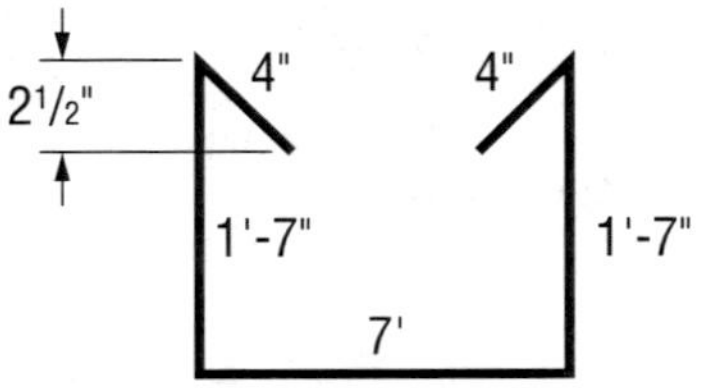
3K2	6	#3	6'3"	14,10	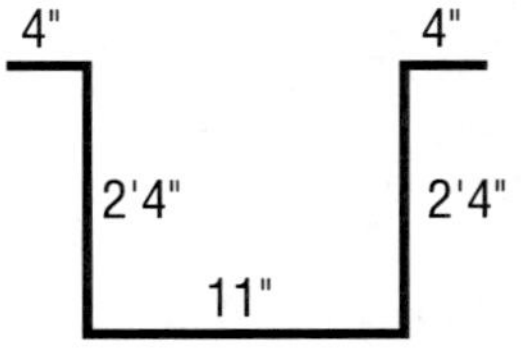
3K3	6	#3	1'7"	3,57	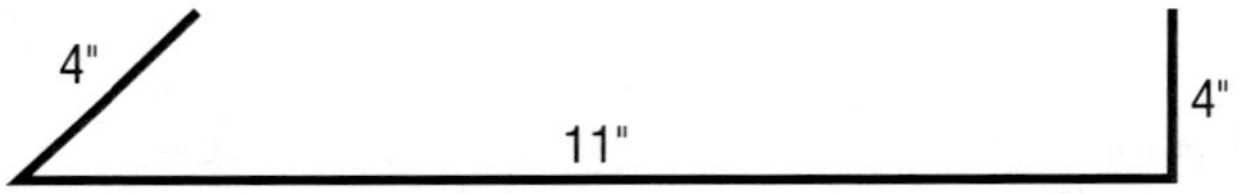
8K6	4	#8	6'10"	72,98	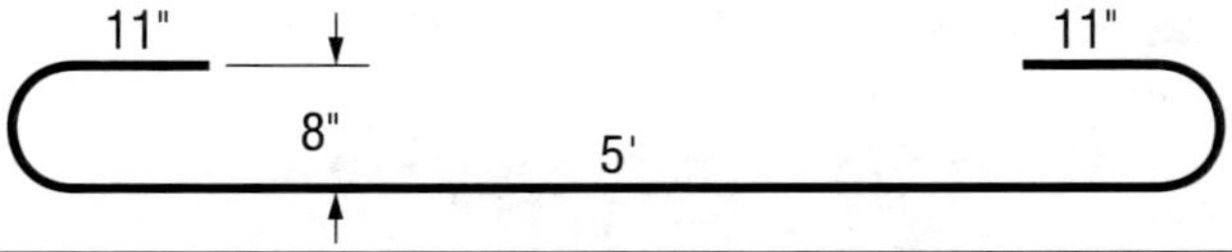
9K1	3	#9	6'7"	67,15	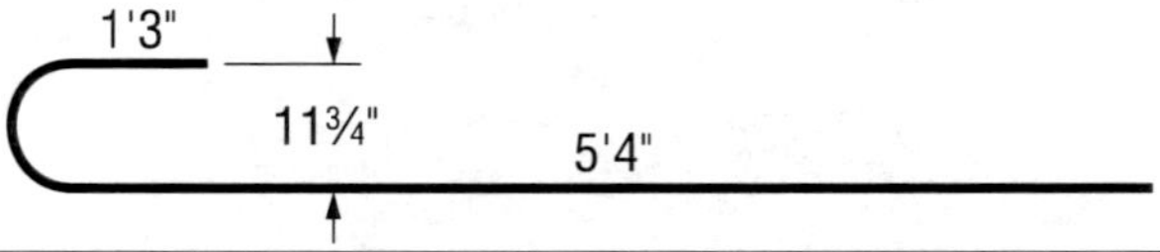
10K4	3	#10	40'6"	522,81	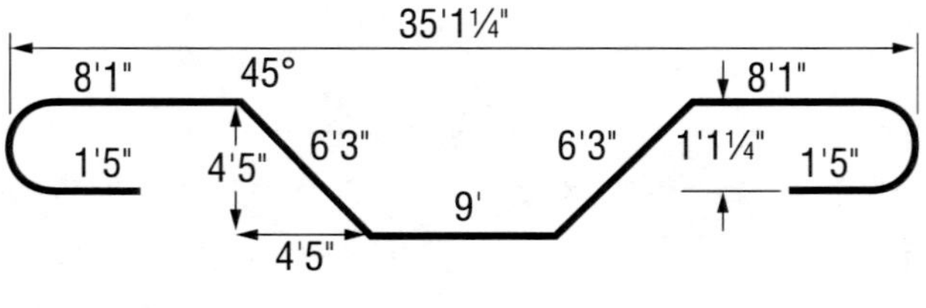

3 = 35,82   5 = 66,58   8 = 72,98   9 = 262,86   10 = 522,81   TOTAL = 951,05

27304-14_F21.EPS

Figure 21 Additional sample bar list.

Client: SAMPLE CLIENT 1   Project: SAMPLE IMPERIAL PROJECT

Address: New York, Elm Street, 67

Code: 2001-1   Number: 1   Module: FACTORY BLDG.

Description: FIRST FLOOR FRAMING PLAN   Type:

Location: FIRST FLOOR   Machine: Label: 3

Row: 6   Mark: 3K2   Bars: 6   #3   Long: 6'3"   Wt: 14,10

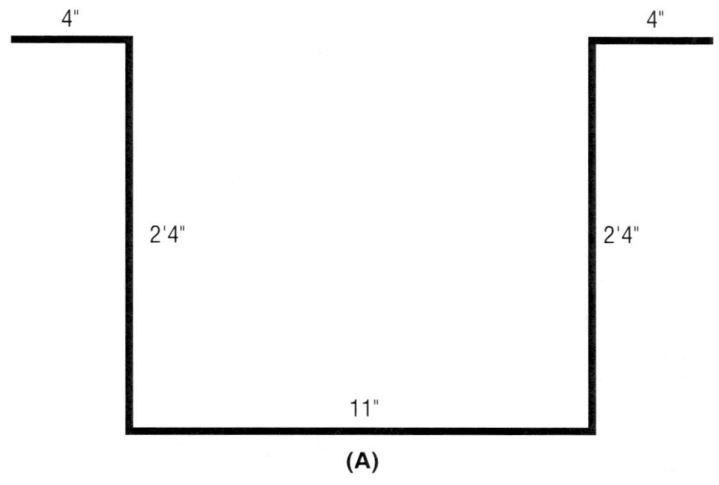

(A)

Client: SAMPLE CLIENT 1   Project: SAMPLE IMPERIAL PROJECT

Address: New York, Elm Street, 67

Code: 2001-1   Number: 1   Module: FACTORY BLDG.

Description: FIRST FLOOR FRAMING PLAN   Type:

Location: FIRST FLOOR   Machine: Label: 7

Row: 10   Mark: 10K4   Bars: 3   #10   Long: 40'6"   Wt: 522,81

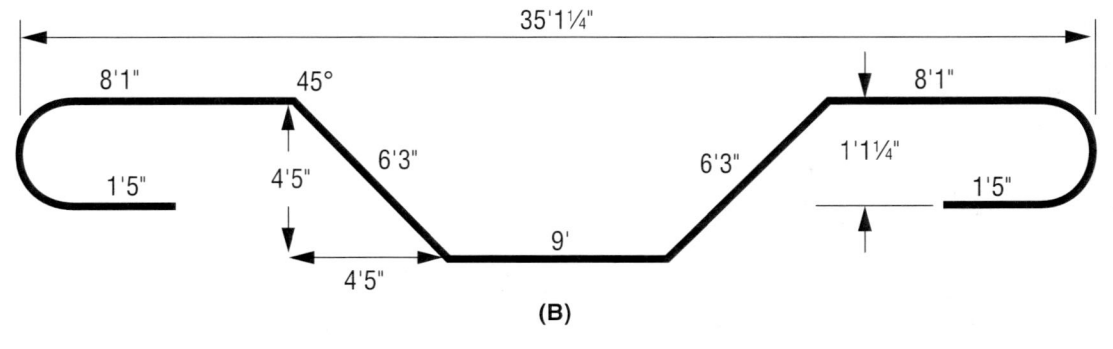

(B)

27304-14_F22.EPS

Figure 22 Examples of rebar labels.

27304-14 **Reinforcing Concrete**   Module Five   21

These classes of steel-wire bar supports can be used in many different configurations. *Figure 23* shows various wire bar-support configurations. Some of the more common supports are as follows:

- *Slab bolster (SB)* – Used under the rebars placed in slabs. The top wire is corrugated, providing firm tying of bars with any spacing. The legs are on 5" centers. Slab bolsters are stocked in heights of ¾" to 2", in lengths of 5' and 10'.
- *Slab bolster upper (SBU)* – Provides support for rebars in slabs on cork or fill. This is essentially a slab bolster with two wires welded lengthwise along the base. SBUs are available in the same sizes as slab bolsters, but usually in Class A only, except on special order.
- *Slab bolster plate (SBP)* – Similar in design to slab bolster supports except that steel plates, instead of wires, are welded lengthwise along the base. The applications and sizes are the same as those for the slab bolster uppers. They are also called sand plates.
- *Beam bolster (BB)* – Supports rebars placed in beams and girders regardless of the spacing. The legs are spaced 2½" apart. They are available in 5' lengths and may be cut to the particular beam widths needed on the job. BBs are available in different heights.
- *Beam bolster upper (BBU)* – Provides support between layers of bars. Two wires are welded lengthwise along the base of a beam bolster. The heights and lengths are the same as those for the beam bolster. BBUs are available in Class A only, except on special order.
- *Bar chair (BC)* – Supports miscellaneous reinforcing steel and may, in certain applications, be used as a substitute for slab bolsters. BCs are available in different heights.
- *Joist chair (JC)* – Supports rebars placed in concrete ribs or joists of metal-pan or cinder-fill concrete slabs. JCs are available in different heights and 4", 5", or 6" widths.
- *High chair (HC)* – Formed from steel wire. It has upturned legs. HCs are available in different heights.

- *Continuous high chair (CHC)* – Supports top slab bars and truss bars at beams, girders, and columns. Available heights are the same as those for the individual high chairs.
- *Continuous high chair upper (CHCU)* – A continuous high chair with two wires welded lengthwise along the base. This supports the upper layers of reinforcing steel and may also be used on fill or cork. CHCUs are available in 5' lengths in the heights available for the individual high chairs. Except on special order, only Class A is available.
- *Joist chair upper (JCU)* – Supports reinforcing steel in pan-joist or waffle-dome work. The supporting bar is either #4 reinforcing bar or ½" plain round bar. Standard width is 14". JCUs are available in heights up to 3½" in Class A only with upturned or end-bearing legs.

2.4.2 *Precast Concrete Blocks*

Precast concrete blocks, also known as dobies, may also be used as bar supports, especially in footings where the bars usually clear the bottom by 3".

Blocks should be placed about 5' off center each way under the footing mat. At least four blocks should be used per mat. There are a number of precast concrete bar supports in use, as illustrated in *Figure 24*. The most common are the plain- and wired-block types. Any blocks used should have the same compressive strength as the concrete being placed.

2.4.3 *Other Types of Bar Supports*

A number of all-plastic supports are used in place of steel or concrete when corrosion must be avoided or appearance is a factor. Some of these plastic supports are shown in *Figure 25*.

A steel standee (*Figure 26*) may also be used as a bar support. The standee is a U-shaped support having two legs bent at 90-degree angles in opposite directions. It is used as a high chair and placed on a lower mat of bars to support an upper mat of bars.

Reinforcing-Bar Sizes

There are 11 standard sizes of manufactured reinforcing bars. These are identified by a number expressed in inch-pound units ranging from #3 to #18, or in metric units with sizes ranging from #10 to #57. In either system, the size number represents the approximate diameter of the bar. In the inch-pound system, the size number represents eighths of an inch; in the metric system it represents millimeters (mm). For example, a #5 bar has an approximate diameter of ⅝". When expressed in metric system units, the corresponding #16 bar has an approximate size of 16 mm (about ⅝").

BAR SUPPORT ILLUSTRATION	BAR PLASTIC ILLUSTRATION PLASTIC CAPPED OR DIPPED	SYMBOL	TYPE OF SUPPORT	TYPICAL SIZES
5"	CAPPED 5"	SB	Slab Bolster	$\frac{3}{4}$, 1, $1\frac{1}{2}$, and 2 inch heights in 5 ft and 10 ft lengths
5"		SBU* SBP ↓	Slab Bolster Upper or Slab Bolster Plate	Same as SB
$2\frac{1}{2}$" $2\frac{1}{2}$"	CAPPED $2\frac{1}{2}$" $2\frac{1}{2}$"	BB	Beam Bolster	1, $1\frac{1}{2}$, 2 to 5 inch heights in increments of $\frac{1}{4}$ inch in lengths of 5 ft
$2\frac{1}{2}$" $2\frac{1}{2}$"		BBU*	Beam Bolster Upper	Same as BB
	DIPPED	BC	Individual Bar Chair	$\frac{3}{4}$, 1, $1\frac{1}{2}$, and $1\frac{3}{4}$ inch heights
DIPPED	DIPPED	JC	Joist Chair	4, 5, and 6 inch widths and $\frac{3}{4}$, 1, and $1\frac{1}{2}$ inch heights
	CAPPED	HC	Individual High Chair	2 to 15 inch heights in increments of $\frac{1}{4}$ inch
		HCM*	High Chair for Metal Deck	2 to 15 inch heights in increments of $\frac{1}{4}$ inch
8"	CAPPED 8"	CHC	Continuous High Chair	Same as HC in 5 ft and 10 ft lengths
8"		CHCU*	Continuous High Chair Upper	Same as CHC
		CHCM*	Continuous High Chair for Metal Deck	Up to 5 inch heights in increments of $\frac{1}{4}$ inch
Top of slab #4 OR $\frac{1}{2}$"⌀ Height 14"	Top of slab #4 OR $\frac{1}{2}$"⌀ Height 14" DIPPED	JCU**	Joist Chair Upper	14-inch span and heights range from 1 inch to $3\frac{1}{2}$ inches in $\frac{1}{4}$ inch increments
		CS	Continuous Support	$1\frac{1}{2}$ to 12 inch in increments of $\frac{1}{4}$ inch in lengths of 6 to 8 ft

Note: 1 inch = 25.4 mm
 * Usually available in Class A only, except on special order.
** Usually available in Class A only, with upturned or end bearing legs.
 ↓ Slab Bolster Plate (SBP) has steel plates instead of wires welded lengthwise along the base.

27304-14_F23.EPS

Figure 23 Typical wire bar supports.

BAR SUPPORT ILLUSTRATION	SYMBOL	TYPE OF SUPPORT**	TYPICAL SIZES	DESCRIPTION
	PB	Plain Block	A - ¾" to 6" B - 2" to 6" C - 2" to 48"	Used when placing rebar off grade and formwork. When C dimension exceeds 16", a piece of rebar should be cast inside block.
	WB	Wired Block	A - ¾" to 4" B - 2" to 3" C - 2" to 3"	Generally 16-gauge tie wire is cast in block, commonly used against vertical forms or in positions necessary to secure the block by tying to the rebar.
	TWB	Tapered Wired Block	A - ¾" to 3" B - ¾" to 2½" C - 1¼" to 3"	Generally 16-gauge tie wire is cast in block, commonly used where minimal form contact is desired.
	CB	Combination Block	A - 2" to 4" B - 2" to 4" C - 2" to 4" D - fits #3 to #5 bar	Commonly used on horizontal work.
	DB	Dowel Block	A - 3" B - 3" to 5" C - 3" to 5" D - hole to accommodate a #4 bar	Used to support top mat from dowel placed in hole. Block can also be used to support bottom mat.
	DSSS	Side Spacer – Wired	Concrete cover, 2" to 6"	Used to align the rebar cage in a drilled shaft.* Commonly 16-gauge tie wires are cast in spacer. Items for 5" to 6" cover have 9-gauge tie wires at top and bottom of spacer.
	DSBB	Bottom Bolster – Wired	Concrete cover, 3" to 6"	Used to keep the rebar cage off of the floor of the drilled shaft.* Item for 6" cover is actually 8" in height with a 2" shaft cast in the top of the bolster to hold the vertical bar.
	DSWS	Side Spacer for Drilled Shaft Applications	Concrete cover, 3" to 6"	Generally used to align rebar in a drilled shaft. Commonly manufactured with two sets of 16-gauge annealed wires, assuring proper clearance from the shaft wall surface.

* Also known as pier, caisson, or cast-in-drilled hole.

** Blocks should be the same compressive strength as the concrete being placed.

27304-14_F24.EPS

Figure 24 Typical precast concrete bar supports.

2.4.4 *Identification of Bar Supports*

Specifications, drawings, and details identify bar supports by listing their nominal height, length, type, and class of protection. For example, 6 × 5-CHC-A would signify a Class A continuous high chair having a height of 6" and a length of 5'.

2.4.5 *Miscellaneous Accessories*

Figure 27 shows a standard roll of tie wire and a wire tie (pigtail) used to secure lengths of reinforcing steel to each other or to various supports.

2.5.0 Welded-Wire Fabric Reinforcement

When reinforcement is required for concrete pavement, parking lots, driveways, or floor slabs, welded-wire fabric reinforcement (shown as WWF or WWR on construction drawings) can be used instead of individual rebars. Welded-wire fabric reinforcement consists of longitudinal and transverse steel wires electrically welded together to form a square or rectangular mesh or mat. Depending on the wire diameter, which can range up to ¾" or more, welded-wire fabric reinforcement is available in roll form or in 8' × 16' flat

BAR SUPPORT ILLUSTRATION	SYMBOL	TYPE OF SUPPORT	TYPICAL SIZES	DESCRIPTION
	BS	Bottom Spacer	Heights $\frac{3}{4}$" to 6"	Generally for horizontal work. Not recommended for ground or exposed aggregate finish.
	BS-CL	Bottom Spacer	Heights $\frac{3}{4}$" to 2"	Generally for horizontal work; provides bar clamping action. Not recommended for ground or exposed aggregate finish.
	HC	High Chair	Heights $\frac{3}{4}$" to 5"	For use on slabs or panels.
	HC-V	High Chair, Variable	Heights $2\frac{1}{2}$" to $6\frac{1}{4}$"	For horizontal and vertical work. Provides for different heights.
	WS	Wheel Spacer	Concrete Cover $\frac{3}{8}$" to 3"	Generally for vertical work. Bar clamping action and minimum contact with forms. Applicable for concrete-reinforcing steel.
	DSWS	Side Spacer for Drilled-Shaft Applications	Concrete Cover $2\frac{1}{2}$" to 6"	Generally used to align rebar in a drilled shaft.* Two-piece wheel that closes and locks onto the stirrup or spiral, assuring proper clearance from the shaft wall surface.
	VLWS	Locking Wheel Spacer for All Vertical Applications	Concrete Cover $\frac{3}{4}$" to 6"	Generally used in both drilled shaft and vertical applications where excessive loading occurs. Surface spines provide minimal contact while maintaining required tolerance.

* Also known as pier, caisson, or cast-in-drilled hole.

27304-14_F25.EPS

Figure 25 Typical all-plastic supports.

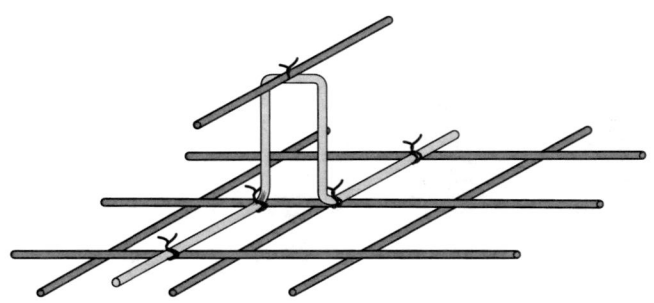

27304-14_F26.EPS

Figure 26 Standee support.

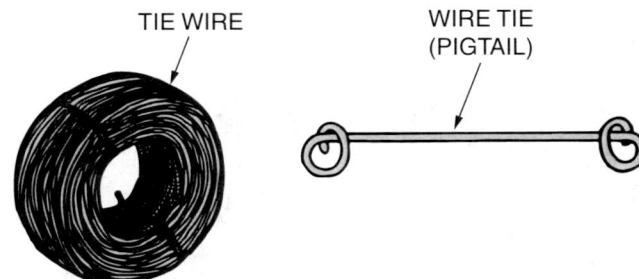

27304-14_F27.EPS

Figure 27 Miscellaneous accessories.

mats. *Figure 28* illustrates some standard sizes of plain-wire welded-wire fabric reinforcement in roll form. Bar supports can be used to space and secure welded-wire fabric reinforcement as well as rebar.

2.5.1 Plain-Wire Reinforcement

Plain wire for welded-wire fabric reinforcement is produced in accordance with *ASTM A82* and is designated by a size number consisting of a W (or MW for metric) followed by the nominal cross-sectional area of the wire in hundredths of a square inch (or square millimeters for metric).

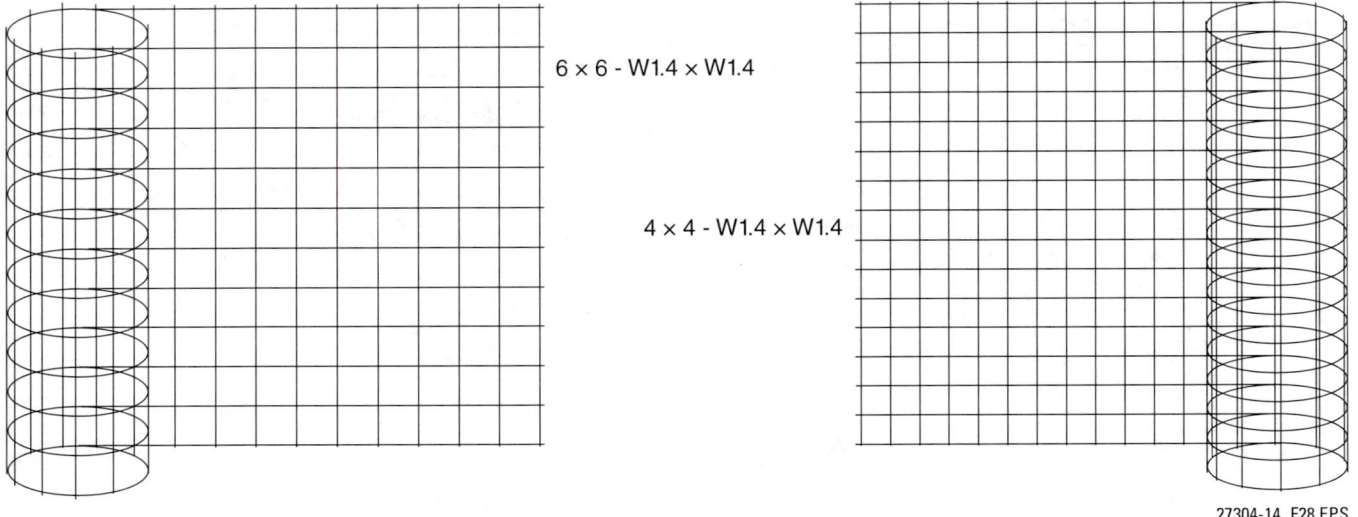

6 × 6 - W1.4 × W1.4

4 × 4 - W1.4 × W1.4

27304-14_F28.EPS

Figure 28 Typical plain-wire, welded-wire fabric reinforcement in roll form.

The wire is manufactured in sizes from W-0.5 to W-45 (MW-3 to MW-290 for metric), as shown in *Table 6*.

Welded-wire fabric reinforcement is designated by a style code such as 4 × 6 – W-10 × W-6 which, in this case, indicates the longitudinal wires are size W-10 on 4" centers and the transverse wires are W-6 on 6" centers. Any style where the transverse wire provides the minimum steel size necessary for fabrication and handling is called one-way reinforcement. Where adequate reinforcement is provided in both transverse and longitudinal directions, the reinforcement is called two-way reinforcement.

Common styles of plain welded-wire fabric reinforcement are shown in *Table 7*. The table also shows the old way of designating welded-wire fabric reinforcement using American Wire Gauge (AWG) wire gauge sizes.

2.5.2 Deformed Welded-Wire Fabric Reinforcement

Deformed wires, manufactured to *ASTM A496*, are also used in making welded-wire fabric reinforcement. The deformations along the wires are intended to improve bonding of the concrete for better crack control. The wires are manufactured and designated in a range from D-1 to D-45 (MD-6 to MD-290 for metric), as shown in *Table 8*. Deformed welded-wire fabric reinforcement is produced to *ASTM A497* and is designated the same as plain welded-wire fabric reinforcement except that the W or MW is replaced by a D or MD. Using the example in the previous paragraph, 4 × 6 – W-10 × W-6 would be 4 × 6 – D-10 × D-6 if deformed wire were used in place of plain wire.

Table 6 Plain Wire Sizes (*ASTM A82*)

Size	Diameter (in)	Area (in²)	Weight (lb/ft)
W-0.5	0.080	0.005	0.017
W-1	0.113	0.010	0.034
W-1.4	0.135	0.014	0.048
W-1.5	0.138	0.015	0.051
W-2	0.159	0.020	0.068
W-2.5	0.178	0.025	0.085
W-2.9	0.192	0.029	0.096
W-3	0.195	0.030	0.102
W-3.5	0.211	0.035	0.119
W-4	0.225	0.040	0.136
W-4.5	0.240	0.045	0.153
W-5	0.252	0.050	0.170
W-5.5	0.264	0.055	0.187
W-6	0.276	0.060	0.204
W-7	0.298	0.070	0.238
W-8	0.319	0.080	0.272
W-10	0.356	0.100	0.340
W-12	0.390	0.120	0.408
W-14	0.422	0.140	0.476
W-16	0.451	0.160	0.544
W-18	0.478	0.180	0.612
W-20	0.504	0.200	0.680
W-22	0.529	0.220	0.748
W-24	0.553	0.240	0.816
W-26	0.575	0.260	0.884
W-28	0.597	0.280	0.952
W-30	0.618	0.300	1.020
W-31	0.628	0.310	1.054
W-45	0.757	0.450	1.531

Table 7 Common Styles of Plain Welded-Wire Fabric

Style Designation		Longitudinal or Transverse Steel Area (in^2/ft)	Approximate Total Weight (lbs/100 ft^2)
Current Designation (by W-number)	**Previous Designation (by steel wire gauge)**		
Rolls			
$6 \times 6 \times W\text{-}1.4 \times W\text{-}1.4$	$6 \times 6 \times 10 \times 10$	0.028	19
$6 \times 6 \times W\text{-}2.0 \times W\text{-}2.0$	$6 \times 6 \times 8 \times 8*$	0.040	27
$6 \times 6 \times W\text{-}2.9 \times W\text{-}2.9$	$6 \times 6 \times 6 \times 6$	0.058	39
$6 \times 6 \times W\text{-}4.0 \times W\text{-}4.0$	$6 \times 6 \times 4 \times 4$	0.080	54
$4 \times 4 \times W\text{-}1.4 \times W\text{-}1.4$	$4 \times 4 \times 10 \times 10$	0.042	29
$4 \times 4 \times W\text{-}2.0 \times W\text{-}2.0$	$4 \times 4 \times 8 \times 8*$	0.060	41
$4 \times 4 \times W\text{-}2.9 \times W\text{-}2.9$	$4 \times 4 \times 6 \times 6$	0.087	59
$4 \times 4 \times W\text{-}4.0 \times W\text{-}4.0$	$4 \times 4 \times 4 \times 4$	0.120	82
Sheets			
$6 \times 6 \times W\text{-}2.9 \times W\text{-}2.9$	$6 \times 6 \times 6 \times 6$	0.058	39
$6 \times 6 \times W\text{-}4.0 \times W\text{-}4.0$	$6 \times 6 \times 4 \times 4$	0.080	54
$6 \times 6 \times W\text{-}5.5 \times W\text{-}5.5$	$6 \times 6 \times 2 \times 2^\dagger$	0.110	75
$4 \times 6 \times W\text{-}4.0 \times W\text{-}4.0$	$4 \times 4 \times 4 \times 4$	0.120	82

*Exact W-number size for eight gauge is W-2.1
†Exact W-number size for two gauge is W-5.4

Table 8 Deformed Wire Sizes (*ASTM A496*)

Size	Diameter (in)	Area (in^2)	Weight (lbs/ft^2)	Size	Diameter (in)	Area (in^2)	Weight (lbs/ft^2)
D-1	0.113	0.01	0.034	D-17	0.465	0.17	0.578
D-2	0.159	0.02	0.068	D-18	0.478	0.18	0.612
D-3	0.195	0.03	0.102	D-19	0.491	0.19	0.646
D-4	0.225	0.04	0.136	D-20	0.504	0.20	0.680
D-5	0.252	0.05	0.170	D-21	0.517	0.21	0.714
D-6	0.276	0.06	0.204	D-22	0.529	0.22	0.748
D-7	0.298	0.07	0.238	D-23	0.541	0.23	0.782
D-8	0.319	0.08	0.272	D-24	0.553	0.24	0.816
D-9	0.338	0.09	0.306	D-25	0.564	0.25	0.850
D-10	0.356	0.10	0.340	D-26	0.575	0.26	0.884
D-11	0.374	0.11	0.374	D-27	0.586	0.27	0.918
D-12	0.390	0.12	0.408	D-28	0.597	0.28	0.952
D-13	0.406	0.13	0.442	D-29	0.608	0.29	0.986
D-14	0.422	0.14	0.476	D-30	0.618	0.30	1.020
D-15	0.437	0.15	0.510	D-31	0.628	0.31	1.054
D-16	0.451	0.16	0.544	D-45	0.757	0.45	1.531

Synthetic-Fiber Reinforcement

Among the concerns associated with welded-wire fabric reinforcement are that it is labor intensive to install and often gets bent out of position by workers walking on it. In recent years, there has been a trend to replace welded-wire fabric reinforcement with fibers made of synthetic material in slabs-on-grade and elevated decks. These fibers are added to the concrete mix by the ready-mix supplier. The fiber additives have proven effective in reducing shrinkage and controlling cracking. In some applications, steel fibers are added to the concrete mix, alone or in combination with synthetic fibers. The use of steel fibers is said to help improve the structural strength and impact resistance of concrete.

Additional Resources

29 *CFR* 1926, *Safety and Health Regulations for Construction*, latest edition. Washington, D.C.: Occupational Safety and Health Administration.

ACI 315, Details and Detailing of Concrete Reinforcement, Latest Edition. Farmington Hills, MI: American Concrete Institute.

ACI 318-95, Building Code Requirements for Structural Concrete, Latest Edition. Farmington Hills, MI: American Concrete Institute.

ASTM A615, Standard Specification for Deformed and Plain Carbon-Steel Bars for Concrete Reinforcement, Latest Edition. West Conshohocken, PA: ASTM International.

ASTM A706, Standard Specification for Low-Alloy Steel Deformed Bars and Plain Bars for Concrete Reinforcement, Latest Edition. West Conshohocken, PA: ASTM International.

ASTM A996, Standard Specification for Rail-Steel and Axle-Steel Deformed Bars for Concrete Reinforcement, Latest Edition. West Conshohocken, PA: ASTM International.

2.0.0 Section Review

1. ACI specifies that reinforcing spirals and ties in columns must have a concrete cover of _____.

   a. 3"
   b. 2½"
   c. 2"
   d. 1½"

2. After each tie is made, the loose ends of tie wires must be _____.

   a. bent down
   b. covered with tape
   c. cut off
   d. wrapped around the rebar

3. The standard configuration for reinforcing bar is the _____.

   a. perforated bar
   b. deformed bar
   c. ridged bar
   d. plain bar

4. The U-shaped bar support that is used as a high chair on a lower mat of bars to support an upper mat of bars is called a _____.

   a. tie wire
   b. spreader
   c. bulkhead
   d. standee

5. When specifying welded-wire fabric reinforcement, such as 4 × 6 – D-10 × D-6, the letter D indicates _____.

   a. deformed wire
   b. plain wire
   c. driven rebar
   d. wire depth

SECTION THREE

3.0.0 BENDING AND CUTTING REINFORCING STEEL

Objective

Describe methods by which reinforcing bars may be bent and cut in the field.

    a. Describe how to cut rebar.
    b. Describe how to bend rebar.

Performance Task

Use appropriate tools to cut and bend reinforcing bars.

The two main field tasks associated with rebar are cutting and bending the material to the correct specifications. Although rebar is commonly fabricated off site, some field fabrication may also be required.

3.1.0 Cutting Rebar

Rebar can be cut to close tolerances using a portable band saw or abrasive (chop) saw. These methods are suitable for those situations that require a bar to be cut to close specifications.

Manual rebar cutters (*Figure 29*) may also be used to cut rebar and are available in various sizes. Rods up to ½" in diameter may be cut quickly and easily using manual rebar cutters.

27304-14_F29.EPS

Figure 29 Manual rebar cutter.

Manual or portable power shears (*Figure 30*) are also used to cut rebar, especially portable power shears on those job sites where a great deal of cutting must be done. These machines are available in various sizes and require special care during use. Only personnel who have been properly trained should use power equipment.

3.2.0 Bending Rebar

Reinforcing bars should be bent cold. Rebar may be bent by hand using a tool called a hickey bar or another type of rebar bender (see *Figure 31*). Hickey bars are not recommended for field bending of rebar larger than #5. *Figure 32* illustrates two types of hickey bars: straight and offset. For additional information on bending rebar in the field, refer to the latest publication of the CRSI.

Jigs may also be set up on a table. *Figure 33* shows a simple bending jig made of angle iron welded to a flat plate. The plate is then bolted securely to a table or other stable surface. Jigs may be used singly or in combination to fabricate column ties or bars requiring special bends.

Fabrication shops use power benders (*Figures 34* and *35*). There are portable models available that may be found on some job sites. Power benders are usually able to bend any size rebar to any desired shape. No preheating is necessary when bending the larger sizes of bars with a power bender. Only workers trained in their use should use a power bender.

Rebar Cutter

A portable rebar cutter like the one shown here can be used to cut rebar up to and through size #6 (¾").

CUTTER

27304-14_SA04.EPS

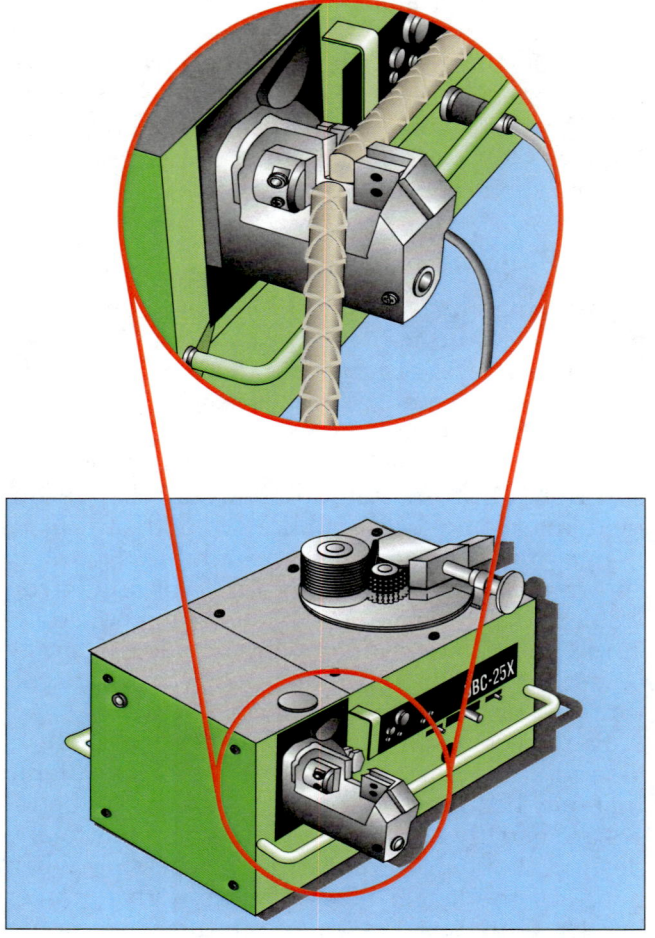

27304-14_F30.EPS

Figure 30 Rebar being sheared to length by a portable power shears.

HICKEY BAR　　　　**REBAR BENDER**

27304-14_F31.EPS

Figure 31 Hickey bar and rebar bender.

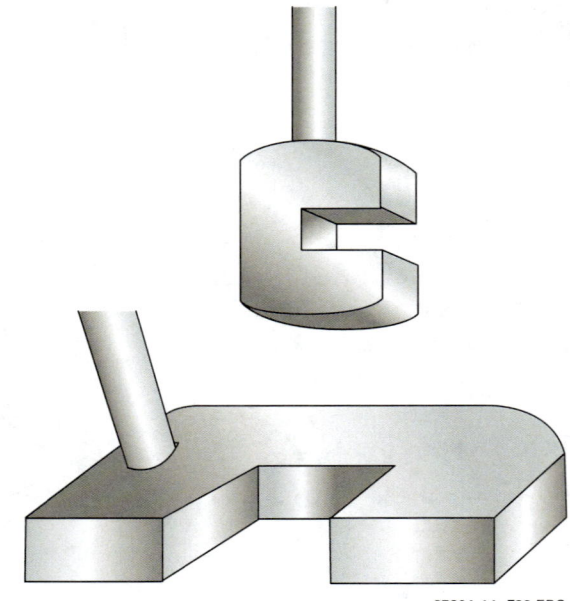

27304-14_F32.EPS

Figure 32 Fabricated hickey bars.

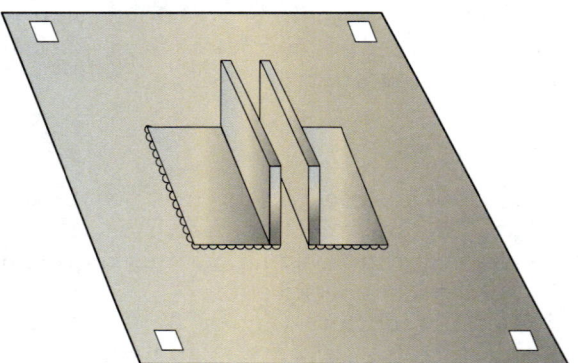

27304-14_F33.EPS

Figure 33 Bending jig.

27304-14_F34.EPS

Figure 34 Bending reinforcing bars.

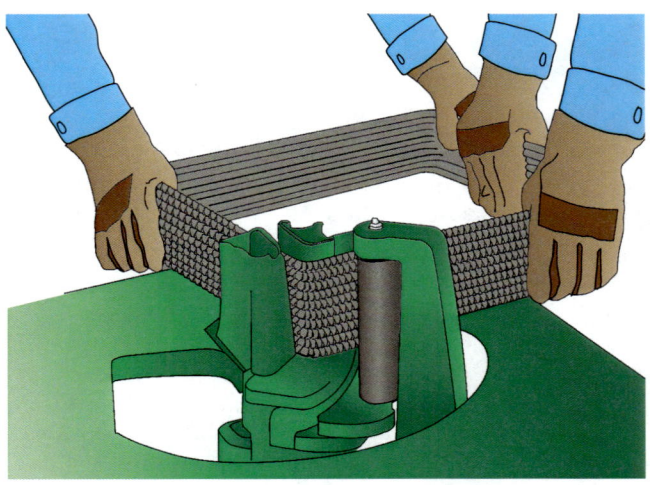

27304-14_F35.EPS

Figure 35 Multiple bending of closed column ties on a power stirrup bender.

3.0.0 Section Review

1. Manual rebar cutters are used to cut rebar up to 1 inch in diameter.

   a. True
   b. False

2. When rebar must be bent in the field, rebar smaller than #5 bar can be bent using a manual tool called a _____.

   a. hockey stick
   b. monkey bar
   c. hickey bar
   d. rod wrapper

4.0.0 PLACING REINFORCING STEEL

Objective

Explain the methods for placing reinforcing steel.

a. Discuss the proper method for tying and splicing reinforcing steel.
b. Explain the proper procedure for placing reinforcing steel.

Performance Tasks

Demonstrate five types of ties for reinforcing bars.

Demonstrate proper lap splicing of reinforcing bars using wire ties.

Demonstrate the proper placement, spacing, tying, and support for reinforcing bars.

Trade Terms

Band: Reinforcing steel in columns that is wrapped around the vertical bars to counteract compression forces.

Contact splice: A means of connecting reinforcing bars by lapping in direct contact.

Double-curtain wall: A concrete wall that contains a layer of reinforcement at each face.

Far face: The face farthest from the viewer (as of a wall); may be the outside or inside face, depending on whether one is inside looking out or outside looking in.

Flat slab: A concrete slab reinforced in two or more directions, with drop panels but generally without beams, and with or without column capitals.

Lapped splice: The joining of two reinforcing bars by lapping them side by side, or the length of overlap of two bars; similarly, the side and end overlap of sheets or rolls of welded-wire fabric.

Near face: The face nearest the viewer, which may be inside or outside, depending on whether one is inside looking out or outside looking in.

Pile cap: A structural member placed on the tops of piles and used to distribute loads from the structure to the piles.

Rebar horses: Wood or metal supports that are used in groups of two or more to hold main reinforcing in a convenient position for placing ties while prefabricating column, beam, or pile cages.

Schedule: A table on placing drawings that lists the size, shape, and number of bars each way, and the mark number of the bars if they are bent.

Single-curtain wall: A concrete wall that contains a single layer of vertical or horizontal reinforcing bars in the center of the wall.

Staggered splices: Splices in bars that are not made at the same point.

Strips: Bands of reinforcing bars in flat-slab or flat-plate construction. The column strip is a quarter-panel wide on each side of the column center line and runs from column to column. The middle strip is half a panel in width, filling in between column strips, and runs parallel to the column strips.

Support bars: Bars that rest upon individual high chairs or bar chairs to support top bars in slabs or joists, respectively. They are usually #4 bars and may replace a like number of temperature bars in slabs when properly lap spliced; also used longitudinally in beams to provide support for the tops of stirrups. Also called raiser bars.

Temperature bars: Bars distributed throughout the concrete to minimize cracks due to temperature changes and concrete shrinkage.

Template: A device used to locate and hold dowels, to lay out bolt holes and inserts, etc.

Weephole: A drainage opening in a wall.

When reinforcing steel is properly cut and bent to specifications, it then must be placed into the forms and tied down to prevent the steel from moving during concrete placement. In some cases, shorter pieces of rebar must be spliced together to extend longer distances. This section provides information on the proper tying, splicing, and placing of reinforcing steel.

4.1.0 Tying and Splicing Reinforcing Steel

The wire used for tying rebar is usually 16-gauge black, soft-annealed wire. Galvanized wire is also available. Some applications may require the use of a heavier-gauge wire such as #15 or #14. The lower the number, the thicker the wire. The most common types of ties are shown in *Figure 36*. Each tie has a particular application.

The snap tie is the simplest and most basic of all ties. Several other ties end in a snap tie. This type of tie is normally used in flat, horizontal work to prevent the reinforcing bars from moving during concrete placement. The snap tie is made by wrapping the wire diagonally once around the two crossing bars. This should be done so that the ends of the wire end up on top to facilitate easy twisting. The ends are then twisted together with a pair of pliers until they are very tight against the bars.

The wrap-and-snap tie, or wall tie, is normally used when tying rebars placed in walls to keep the horizontal bars from shifting during concrete placement. The tie is made by wrapping the wire 1½ times around the vertical bar, then diagonally around the horizontal bar, ending in a snap tie.

The saddle tie is often used to hold the hooked ends of bars in position when tying footing or other mats. It is also used to secure column ties to vertical bars. The tie is made by passing the wire halfway around one of the bars (either vertically or horizontally) on each side of the crossing bar. The wires are brought squarely around the crossing bar, then up and around the first bar, where they are twisted.

The wrap-and-saddle tie is used to secure column ties to vertical bars when there might be a strain on the ties. It is sometimes used to secure heavy mats that are to be lifted by crane. The tie is made like the saddle tie with one exception: the

wire is first wrapped 1½ times around the first bar.

The figure-eight tie is sometimes used in walls instead of the wrap-and-snap tie, but because it takes longer to tie, it is not usually recommended.

The nail-head tie (*Figure 37*) is used when nails are employed to hold wall bars away from forms. The tie is made by wrapping the wire once around the nail head, crossing the wire, and then wrapping it around the outside bar of the wall mat. The bar is drawn tight against the nail head by twisting the ends of the wire.

Reinforcing bars are tied to prevent their movement during normal construction processes or concrete placement. Tying adds nothing to the strength of the finished structure. Therefore, it is not necessary to tie rebars at every intersection. In most cases, tying 25, 33, or 50 percent of the intersections is sufficient. However, the perimeter of the mat must be 100 percent tied. When tying bars in mats that are being assembled in place, the size of the reinforcing bars should dictate the percentage of ties. It is an accepted practice to stagger the intersections being tied. This practice gives the mat added rigidity (*Figure 38*).

For pre-assembled mats, enough intersections must be tied to make the mats sufficiently rigid for handling by crane. A general rule of thumb for tying pre-assembled mats is to tie the perimeter 100 percent and the interior 25 percent to 50 percent unless the job specifications require otherwise.

4.1.1 Tying Reinforcing Steel

Since most tying is done in flat, horizontal formwork, learning to tie stiff-legged instead of in a squatting position will eliminate many aches and pains. After finishing a tie, if the wire from the coil is left bent, the next tie will be easier to make and less likely to cause an eye injury.

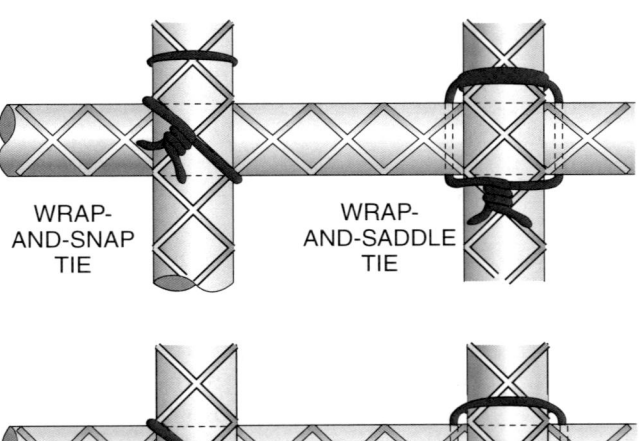
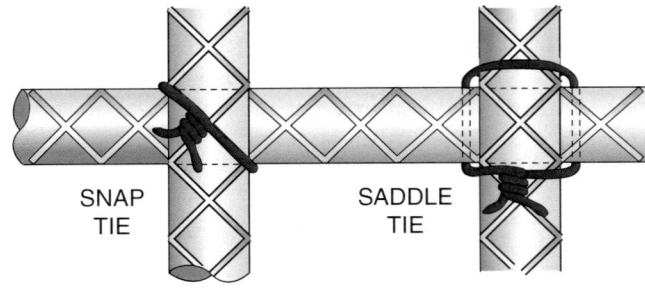
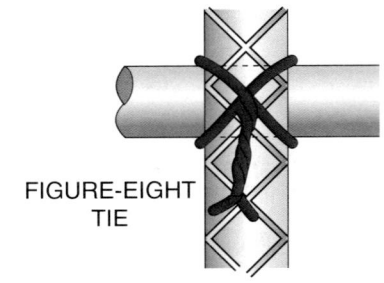

27304-14_F36.EPS

Figure 36 Types of ties.

27304-14_F37.EPS

Figure 37 Nail-head tie.

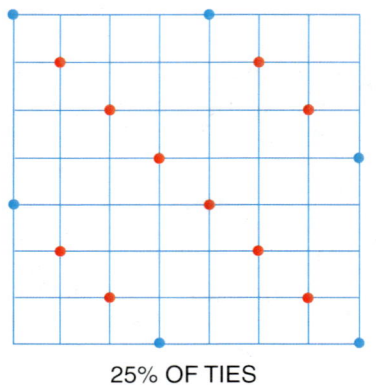
25% OF TIES

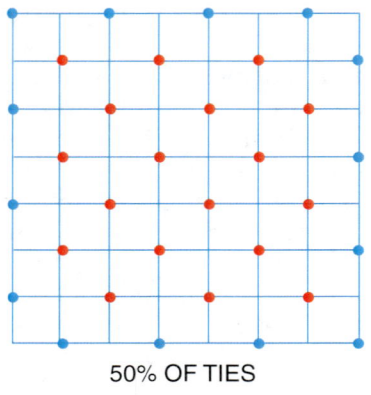
50% OF TIES

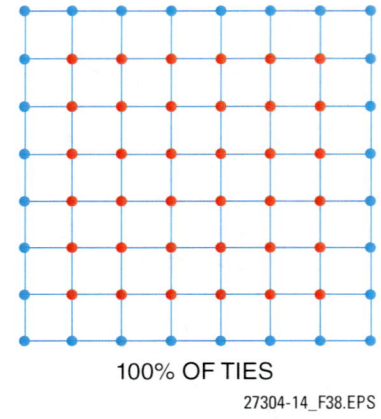
100% OF TIES

27304-14_F38.EPS

Figure 38 Percentage of ties.

To be used effectively, pliers must become an extension of your hand (*Figure 39*). Practice makes perfect in this regard. Develop a rhythm to your work. Wasted motion increases fatigue. In any trade, the true masters are those who make their work look effortless through no waste of motion.

4.1.2 Splicing Reinforcing Steel

Because in most situations it is impossible to provide full-length bars that run continuously throughout a structure, splicing reinforcing bars is common. The placing drawings will show the location and type of splice to use. Sometimes several methods of splicing will be listed by the placing drawings, and the best method may be chosen from among them. No splice should be used, however, without consulting the proper authority.

There are three basic types of splices used in reinforcing steel work. These are lapped splices, welded splices, and mechanical-coupling splices.

As suggested by the name, a lapped splice joins two pieces of reinforcing steel by placing them side by side. In general, three variables affect the length of the lap. These are:

- *Strength of the concrete* – The stronger the concrete, the shorter the splices.

- *Grade of reinforcing steel* – The higher the grade, the shorter the splices.
- *Size of the bars* – The larger the bar, the longer the splices.

These are general statements that have been included to provide an understanding of why certain lengths for certain splices are specified. In actual practice, the amount of overlap is determined by a variety of design considerations as governed by *ACI 318*, which contains tables that provide detailed requirements for lap splice lengths. The placing drawings will usually provide all the necessary information concerning the lengths of the splices. A rule of thumb is to provide a lap equal to 30 times the bar diameter or 12", whichever is greater.

There are two kinds of lapped splices. The contact splice shown in *Figure 40* is made by placing the bars next to each other so that they touch (based on the latest ACI specification).

The spaced-lap splice shown in *Figure 41* is made by placing the bars a certain distance from one another without any actual contact. The maximum center-to-center spacing of the bars must not exceed 6".

27304-14_F39.EPS

Figure 39 Pliers used for tying rebar.

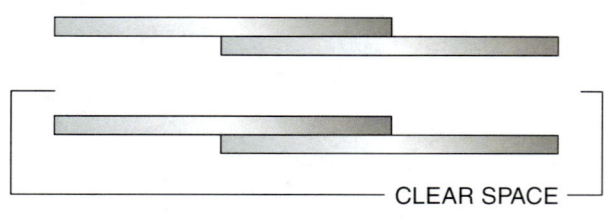

CLEAR SPACE

MIN. FOR BEAMS = REBAR DIAMETER OR 1" WHICHEVER IS LARGER

MIN. FOR COLUMNS = 1.5 x DIAMETER OR 1½", WHICHEVER IS LARGER

27304-14_F40.EPS

Figure 40 Contact splice.

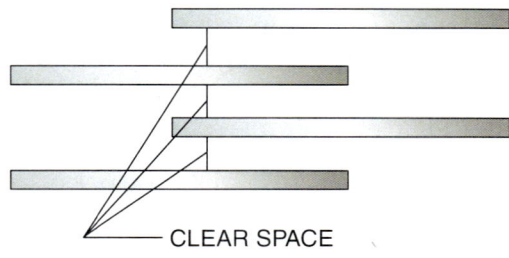

CLEAR SPACE

FOR BEAMS = d OR 1"
WHICHEVER IS LARGER

FOR COLUMNS = 1½d
OR 1½", WHICHEVER IS LARGER

27304-14_F41.EPS

Figure 41 Spaced-lap splice.

Of the two, the contact splice is preferred over the spaced-lap splice because it provides more strength to the finished concrete.

ACI prohibits lap splices of #14 and #18 bars except when these bars are spliced to smaller-sized dowels at footings.

There are basically two types of mechanical splices: coupling splices and end-bearing splices. They are differentiated by the types of forces they resist.

- *Coupling splices* – These splices resist both compression and tension forces. In other words, they are able to withstand forces that try to crush them and pull them apart. One such coupling splice is the forged-steel sleeve. This splice is made by heating a special sleeve until it becomes workable. The ends of the bars are then placed into it, and a portable hydraulic forge crimps the sleeve, forcing the inside of the sleeve against the deformations of the bars. As the sleeve cools, the shrinkage forces cause a

tighter bond. Threaded couplers are also available.
- *End-bearing splices* – A bolted-clamp assembly is available that is used primarily in splicing vertical rebar for columns. The splice is made first by bolting a clamp to the bottom bar. The upper bar is then placed and plumbed, and the upper bolts of the clamp are tightened. Reducers are available that enable a bar of one size to be spliced to a bar of the next smaller size.

4.2.0 Placing Reinforcing Steel

This section provides an overview of the mechanics of placing reinforcing steel.

4.2.1 Placing Bars in Footings and Foundations

A footing is the part of a foundation that rests upon the earth. Footings are placed under columns, piers, and walls to provide the necessary support for these structural members.

The function of a footing is to transmit the concentrated load of a structure to the ground at a pressure that is safe and will not cause settling of the structure. Reinforcing steel is placed within the concrete footing to counteract the forces that tend to bend or break the concrete.

Footings vary greatly in terms of size and shape, depending on the type of load they must carry and the condition of the ground. Generally, the softer the ground, the larger the footing. Location and placement of bar is critical to achieve engineered design strength.

The foundation placing drawings show the individual footings. These drawings usually con-

Rebar Location and Coverage Is Critical

Location and placement of reinforcing bars is critical to ensure that the concrete reaches its design strength and is able to support the intended loads. The following ACI guidelines for rebar placement should be followed unless specific dimensions are provided by the architect or engineer:

- 3" at sides where concrete is cast against earth and on bottoms of footings or other principal structural members where concrete is deposited on the ground
- 2" for bars larger than #5 where concrete surfaces would be exposed to the weather after removal of forms or would be in contact with the ground; 1½" for #5 bars and smaller
- 1½" over spirals and ties in columns
- 1½" to the nearest bars on the top, bottom, and sides of beams and girders
- ¾" for #11 and smaller bars on the top, bottom, and sides of joists, and on the top and bottom of slabs where concrete surfaces are not directly exposed to the ground or weather; 1½" for #14 and #18 bars
- ¾" from the faces of all walls not directly exposed to the ground or weather for #11 and smaller bars; 1½" for #14 and #18 bars

Always refer to the structural notes for clearance and coverage requirements.

tain a **schedule** that lists the size, length, number of bars each way, and mark number of the bars if they are bent. The spacing of the bars may also be provided; if it is not, the bars should be spaced evenly within the footing, based upon the number of bars indicated by the placing drawings. See *Figure 42*.

Very often, bars are assembled into mats prior to installation. All the bars to be placed in one direction are laid out evenly across sawhorses. Then the bars that are located in the opposite direction are laid across the first bars and evenly spaced. The bars are then tied securely at every second or third intersection within the interior of the mat and are usually tied at all intersections around the perimeter to facilitate easier rigging of the mat. Snap ties are normally used for this purpose. If the mat is not to be placed immediately, it should be tagged with the footing number and stored off the ground in a place where it can be easily delivered to the installation site.

The footing mat is usually placed on precast concrete blocks of the proper height. Successive mats are supported in a number of ways depending on the particular needs of the job.

The reinforcing steel for continuous wall footings is normally assembled in place in the trench. Usually, two or more vertical bars are crossed by horizontal bars at spacings indicated by the placing drawings. Dowels, if required, must project above the footing the distance called for by the specifications to allow for the proper vertical splice.

Where two wall footings meet, the horizontal bars may either extend a specified distance into the intersecting footing or be designed with corner bars, depending on the method specified. See *Figure 43*.

A **pile cap** is a structural member placed on the tops of piles. It is used to distribute loads from the structure to the piles. Placing reinforcing steel in pile caps is similar to placing reinforcing steel in footings. The bottom mat rests on individual high chairs placed directly on the piles. Bar placement, however, may not always be at right angles, as shown in *Figure 44*.

Instead of individual column footings or continuous wall footings, a single slab of concrete may be used to support all the units of a structure. The thickness of this slab varies greatly, depending on the soil conditions and loads to be supported. In general, the thickness of the slab is 1' to 5', but it may be as much as 15'.

A mat of reinforcing steel with bars running in two directions is normally used in slabs. The bottom mat generally rests about 3" above subgrade on concrete blocks placed 5' apart in each direction. The top of the mat, or the top layer of bars, is usually about 2" or 3" below the surface of the concrete.

The placing drawings provide all the necessary information concerning the slabs, including the number of pieces, sizes, length, and spacing of the bars. Successive layers of bars are supported by standees or other suitable bar supports.

It is not difficult to determine the spacing of bars within a footing. The first step is to establish the center line of the footing. If an odd number of rebar is to be placed within the footing in one direction, one rebar will sit directly on the center line. If an even number of rebar is to be placed

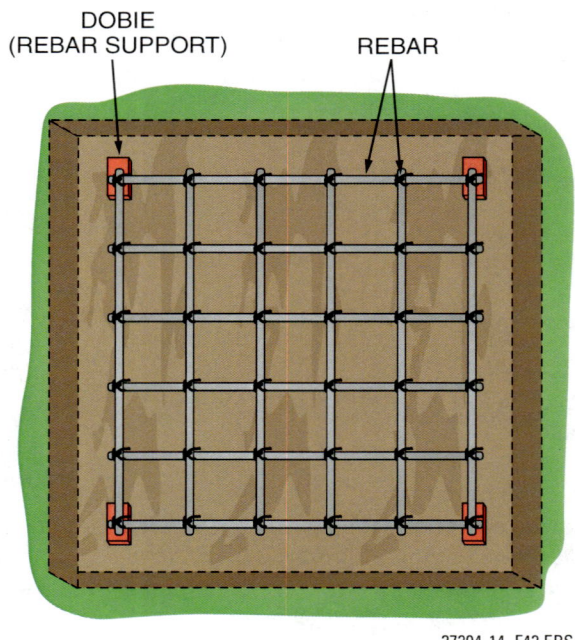

27304-14_F42.EPS

Figure 42 Square footing.

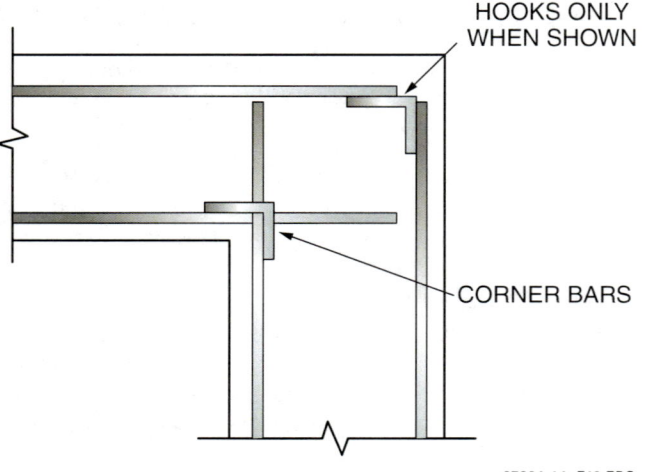

27304-14_F43.EPS

Figure 43 Corner bars.

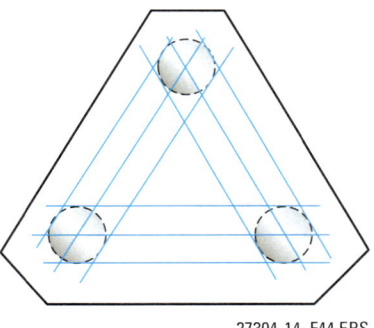

27304-14_F44.EPS

Figure 44 Pile cap.

within the footing, one rebar will lie on either side of the center line. Proceed as follows:

Step 1 Convert all dimensions to inches.

Step 2 Subtract 1 from the number of rebar to be placed in one direction (that is, from those rebar that will lie north-south or east-west).

Step 3 Divide the number of inches by the number of rebar to be placed in one direction minus 1. A remainder indicates the number of inches from each end that a bar will lie.

For example, suppose a mat that is 15' square is called for by the placing drawings. The mat, according to the information found on the placing drawings, is to be made from 22 #5 reinforcing bars, with each bar having a length of 14'-6".

First, the center line should be established and marked. This is true even if the mat is being pre-assembled on sawhorses.

Next, find out where the first bar will be placed in relation to the center line. Since the mat is to be assembled from a total number of 22 bars, the mat will be made of 11 bars running north-south and 11 bars running east-west. Eleven is an odd number; therefore, the first bar will be placed directly on the center line.

Fifteen feet converts to 180". Subtract the center bar from the number of bars to be placed in one direction. This yields 10. Divide 180" by 10; the result is 18". This is the spacing of the bars.

Place the first bar directly on the center line, the next bar 18" away from it (on center), and the next 18" from the previous, etc. Do the same for the bars running in the opposite direction, and tie the bars securely together at as many intersections as needed.

Consider another example. Suppose the same mat is to be made using 24 #5 reinforcing bars instead of 22. First, establish the center line. The mat will be made of 12 bars placed north-south and 12 bars placed east-west. This means that a bar will be placed on each side of the center line.

In this case, 180" is to be divided by 12 minus 1, or 11. This yields 16" with a remainder of 4". This means that two bars will straddle the center line and be 16" apart. (Each of the first two bars will be 8" from the center line.) Place the remaining bars on 16" centers from each other for the rods running in both directions. Tie the mat securely. There will be 4" from the last bar running north-south to the end of the bars running east-west.

4.2.2 Placing Bars in Walls

For the purpose of placing reinforcing steel, each wall has two faces, labeled simply the **far face** and the **near face**, which is a matter of orientation depending on whether the viewer is inside or outside.

Mechanical Rebar Splices

An example of a bolted-clamp assembly used for splicing vertical rebars for larger columns is shown in this figure.

27304-14_SA05.EPS

27304-14 Reinforcing Concrete

The two basic types of reinforced concrete walls are **single-curtain wall** and the **double-curtain wall**. The single-curtain wall is a type of reinforced concrete wall in which a single layer of reinforcing steel is placed in the center between the faces. A double-curtain wall is a reinforced concrete wall in which a layer of reinforcing is placed at each face. A retaining wall, which may be either a single-curtain or double-curtain wall, is used to hold earth or fill in place.

When constructing a reinforced concrete wall, the formwork is usually erected for one face and braced from the outside so that the reinforcing steel may be placed within it. The first step is usually to fasten the vertical bars of the outside face to the dowels projecting upward from the footing. Then, one horizontal bar is typically wired to the vertical bars to keep them plumb. A wrap-and-snap tie is generally used, but a figure-eight tie can be used if more stability is required. The remaining vertical and horizontal bars are then tied at every third intersection, using no fewer than three ties per bar. If the wall being tied is a double-curtain wall, the rebar is tied to the bars of the outside face, and the inside face is set in the same manner. These mats of reinforcing steel can also be constructed on a level surface (prefabricated) and then placed and stabilized in the form.

Reinforcing-steel wall mats must be supported and/or spaced away from the formwork at the top to maintain the proper cover. There are three ways to do this. The first way is to use short lengths

of slab bolsters (as shown in *Figure 45*), beam bolsters, or individual bar chairs. These are stapled to the forms, and the mat is wired to them. Class C, D, or E chairs should be used in areas of exposed concrete to avoid rust.

A second way is to use precast concrete blocks with embedded tie wires. See *Figure 46*. This is the same method used when placing bars in footings. This method is typically used for below-grade applications.

Another method to support wall mats is to drive nails into the forms, leaving them exposed for the required amount of cover, and then tie the

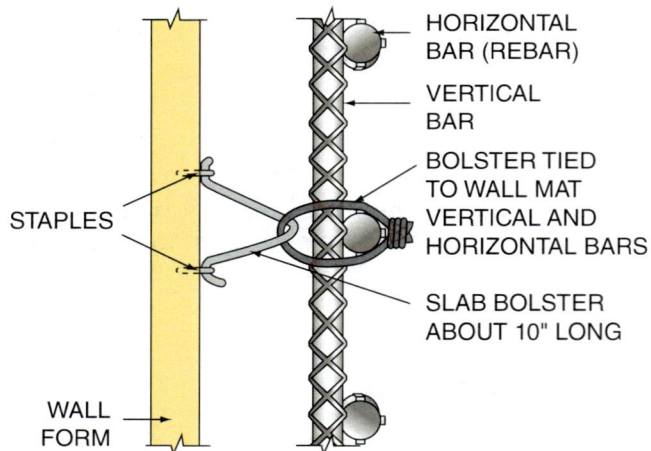

27304-14_F45.EPS

Figure 45 Slab-bolster support.

Automatic Rebar-Tying Tool

This pneumatic tool is designed to wrap, twist, and cut wire in one operation. According to the manufacturer, a person using this tool can tie rebar five times faster than a person manually tying the rebar. Another claimed advantage is the elimination of the repetitive wrist-twisting action that can lead to the condition known as carpal tunnel syndrome. The machine is fed by a spool of wire. Each spool can make more than 100 ties. An extension bar is available so that ties can be made without bending over.

27304-14_SA06.EPS

wall mat to the nails using the nail-head tie (*Figure 47*). This method typically sees limited use; workers can be scratched or cut by an exposed nail and the potential rusting of the nails can be a concern.

In double-curtain walls, the mats must be spaced a specified distance from each other. This information is provided on the placing drawings. One way to space the mats is to use spreader bars. These bars are usually made from #3 rebar and bent into a U or Z shape. They are essentially standees meant to be used in walls instead of slabs.

Prefabricated wire spreaders are also available that both support and spread the wall mats. They are used especially for those situations in which the concrete will be exposed. *Figure 48* illustrates a combination of all the supports and spacers discussed.

When two walls meet at a corner, the bars of the outside faces may be bent 90 degrees or may extend for a specified distance into the intersecting wall and be spliced to a corner bar. The bars of the inside faces usually extend into the intersecting wall but may be provided with hooks. As in all cases, the placing drawings provide all the necessary information. *Figure 49* illustrates the

various methods in use for both single-curtain walls and double-curtain walls.

Reinforcing steel is placed in cantilevered retaining walls in much the same way as it is placed in supported walls, except for the following:

- Bottom-footing reinforcing bars may be bent upward to be used as dowels for the back face of the wall.
- Vertical reinforcing bars are two or three different heights with perhaps every second or third bar extending to the top of the wall. Others terminate at specified cutoff points.
- A **weephole** is usually supplied for drainage. Care must be taken not to alter the position of the pipe when tying reinforcing steel.

4.2.3 *Placing Bars in Columns*

A column is a vertical member used to support a floor beam, girder, or other structural member. The main load to which a column is subjected is compression. To counteract the compression force,

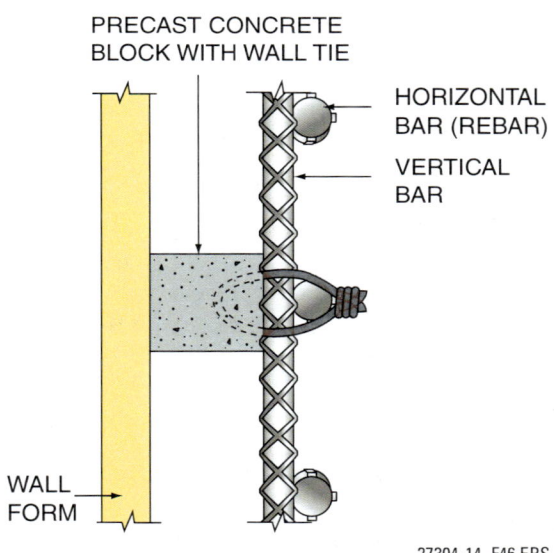

Figure 46 Block support.

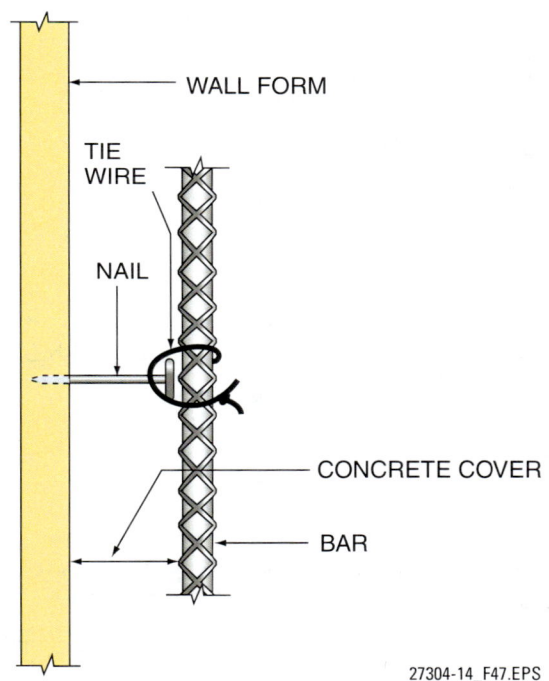

Figure 47 Nail support.

Fabricating Rebar Mats

When it is necessary to fabricate similar rebar mats, you can increase your productivity by making a template. Do this by nailing clean 2 × 4s to the tops of four sawhorses. Then lay out and mark the spacing for the north-south and east-west rebars on the 2 × 4s. Following this, drive nails on both sides of all rebar layout marks. These nails will serve as guides for rod placement and subsequent mat layout and fabrication. This method not only increases your productivity, it also helps maintain accurate spacing of the rebar while they are being placed and tied into position.

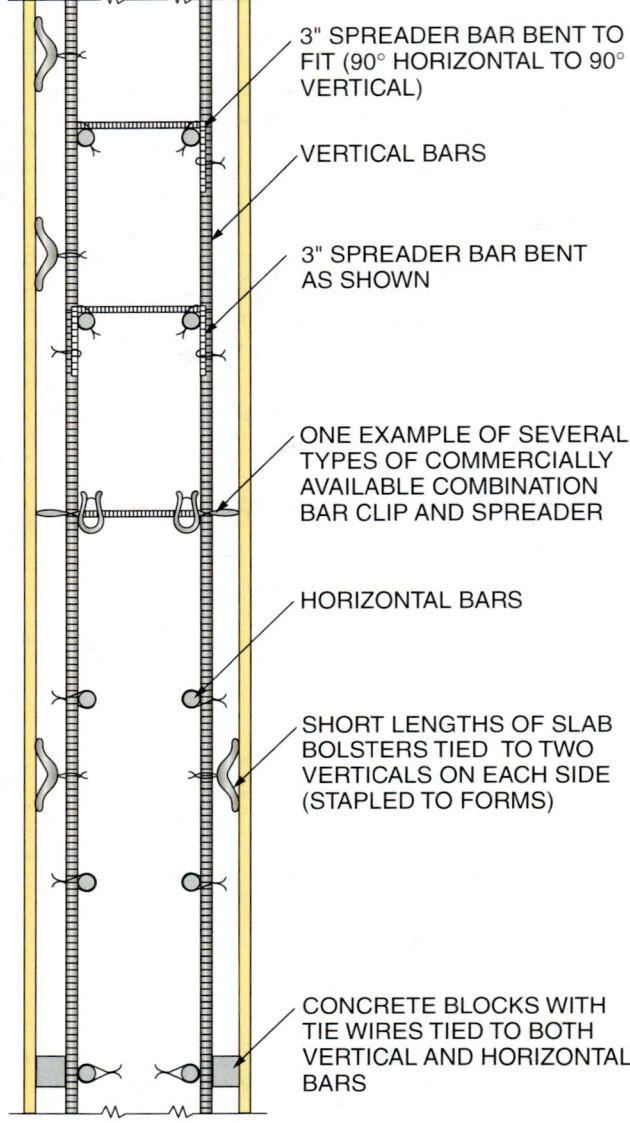

3" SPREADER BAR BENT TO FIT (90° HORIZONTAL TO 90° VERTICAL)

VERTICAL BARS

3" SPREADER BAR BENT AS SHOWN

ONE EXAMPLE OF SEVERAL TYPES OF COMMERCIALLY AVAILABLE COMBINATION BAR CLIP AND SPREADER

HORIZONTAL BARS

SHORT LENGTHS OF SLAB BOLSTERS TIED TO TWO VERTICALS ON EACH SIDE (STAPLED TO FORMS)

CONCRETE BLOCKS WITH TIE WIRES TIED TO BOTH VERTICAL AND HORIZONTAL BARS

27304-14_F48.EPS

Figure 48 Spacers and spreaders.

the reinforcing steel used in columns is wrapped in bands around the vertical bars. This configuration of reinforcing steel, combined with the good resistance concrete exhibits against compression, makes a properly designed and tied column a very strong part of a structure.

Columns that rest on footings are connected to those footings by dowels. A dowel is a steel reinforcing bar that connects two separately cast sections of concrete. Care must be taken to place dowels in the exact places identified in the placing drawings. Dowels must project a specified distance above the footing and be placed accurately with respect to the location of the reinforcing steel

to be placed within the column. Omitting or misplacing dowels can cause serious problems.

A common practice in dowel placement is to use a template (*Figure 50*). A template is made of boards with holes drilled in them to serve as guides for the dowels. Guide boards having a tie or a piece of spiral affixed to them at the location where the dowels are to be placed can also be used (*Figure 51*).

If the dowels have hooks, additional stability can be obtained by tying the hooked end to the footing mat of reinforcing steel.

Dowels should not be pushed into wet concrete. It is very difficult to maintain proper dowel alignment this way, and corrections after the concrete has set are difficult and costly. Whenever possible, templates should be used to obtain the correct positioning.

When placing dowels that are to be butt-spliced to the column vertical bars, extra care is needed because the splices must be staggered. The dowel cage must be constructed with staggered splices in mind, and the placing drawings must be studied very carefully.

Be alert for dowels required for grade beams that join the sides of footings or for dowels that extend into foundation walls or other structural units to be built later.

Another method of placing dowels is to construct a dowel cage, as shown in *Figure 52*. A dowel cage is made by assembling the dowels having 90-degree hooks in such a manner that the hooks provide a stable means of support to the footing mat. The correct placement of the dowels above the footing must be ensured.

Rebar for columns can also be assembled as rebar cages. The bands wrapped around the vertical bars are called column ties. They must be placed outside the vertical bars. These ties can be square, rectangular, U-shaped, circular, or any other shape designated by the engineers. Standards have been developed concerning the placing of column ties. *Figure 53* shows the standard placing of column ties in columns containing an even number of vertical bars. The information in *Figure 53* applies to those cages that are either pre-assembled or erected in place on freestanding, butt-spliced vertical bars.

Figure 54 shows the standard placing of column ties for lap-spliced pre-assembled cages only. The dotted lines indicated in *Figure 54* for the 6-, 8-, and 10-bar columns show how the column ties are to be tied if the distance between the centers of the bars is over 6".

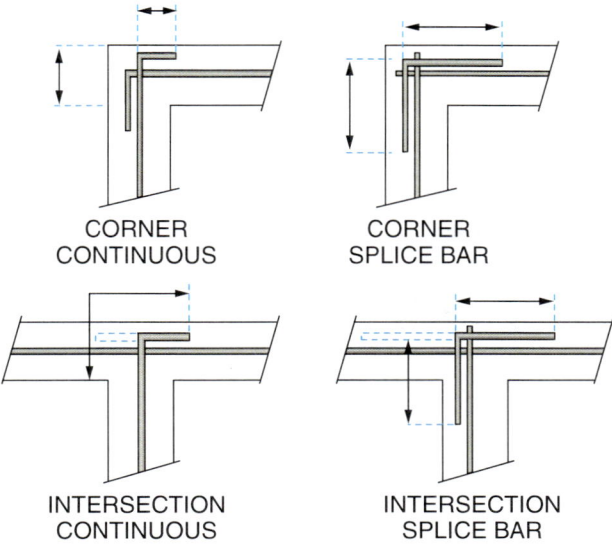

CORNER
CONTINUOUS

CORNER
SPLICE BAR

INTERSECTION
CONTINUOUS

INTERSECTION
SPLICE BAR

INTERSECTION
CONTINUOUS

27304-14_F49.EPS

Figure 49 Corner details.

Each pattern consists of an outside closed tie with pairs of U-shaped bars that are lap-spliced and hooked at each end.

The usual procedure for tying pre-assembled reinforcing column steel is to first lay out all the

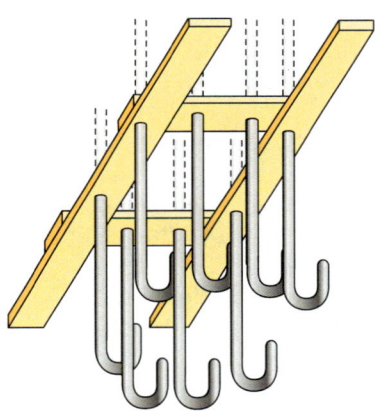

27304-14_F50.EPS

Figure 50 Dowel template.

vertical rebar for one side across supports, called **rebar horses**, as shown in *Figure 55*. Then the vertical column ties are laid out on the vertical rebars and spaced according to the placing drawings. It is generally acceptable to alternate the position of the hooks of the column ties when placing them in sets. The column ties are then wired to the vertical rebar using a saddle tie or a wrap-and-saddle tie for heavy bars. The remaining vertical rebar are put in place and wired to the column ties. When tying columns, it is extremely important to make sure all vertical rebar are lined up and even on the bottom end so that all rebar of the finished column cage will sit firmly on the footing. On 4- and 6-bar columns, every tie should be wired to every vertical rebar at every intersection to achieve good stability. On columns of 6 bars or more, tie 100 percent of the column-corner bars using saddle ties.

Large square or rectangular units may require diagonal wire bracing for greater stability. The bracing should be twisted with pliers until a sufficient amount of tension is obtained.

Cleanouts

When constructing wall or column forms, it is good practice to construct one or more 6" cleanouts at the bottom of the form. The cleanouts provide easy access to the interior of the wall or column forms and allow debris to be removed before concrete is placed. Cleanouts are typically located at the corners and intersections of the form. Always ensure the cleanouts are closed and properly braced before concrete is placed in the forms.

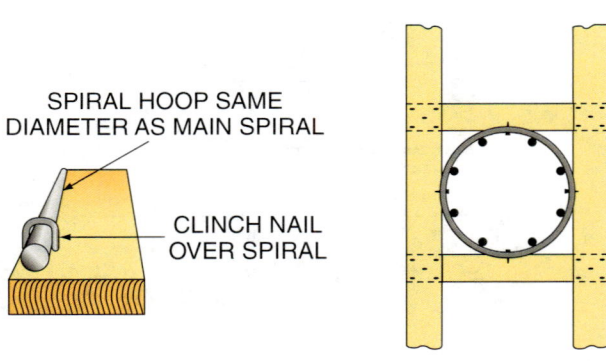

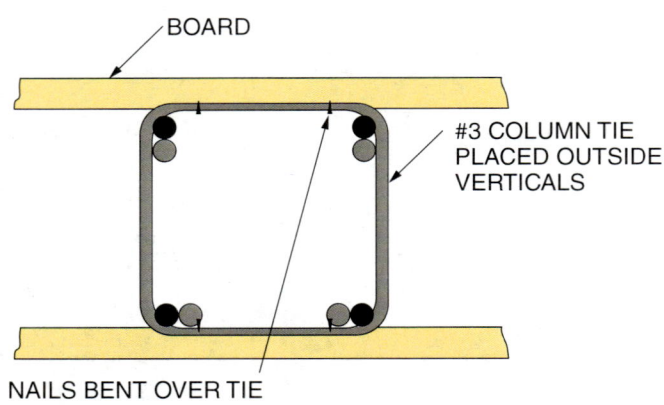

27304-14_F51.EPS

Figure 51 Guide boards.

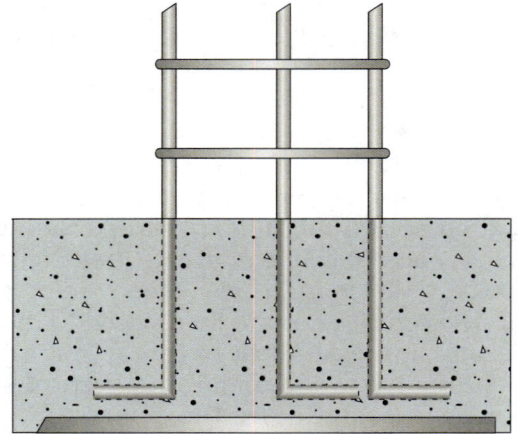

27304-14_F52.EPS

Figure 52 Dowel cage.

Spirals are made of plain rebar or wire shaped like a coil spring. They have spacers attached on opposite sides that keep the turns of the spiral at equal distances. In terms of strength, a column using spirals is generally stronger than a column using square or rectangular column ties.

Spirals are usually shipped collapsed and must be opened or knocked down at the job site prior to assembly. To pre-assemble spirals, two vertical bars are placed inside the spiral and supported at either end by rebar horses. See *Figure 56*. After two bars are placed, the remaining number of vertical rebar specified in the placing drawings are inserted and spaced according to the specifications. The bars are then wired to the spiral to achieve the necessary stability and rigidity.

Column forms should be supported at three or four points, if possible. The supports should be placed as near to the bottom, midsection, and top of the column as possible. Column forms may be supported with nails driven into the inside of the forms in much the same way nails are used to support wall reinforcing. However, precast con-

crete blocks with embedded tie wires are more commonly used.

If precast blocks are used, they should be wired to the column at intersections of the column ties and vertical bars. This will prevent the blocks from spinning when the column is lowered into place within the form.

4.2.4 *Placing Bars in Beams and Girders*

There is no standard sequence that can be listed concerning the placing of bars in beams and girders due to the individual differences in beam height and bar arrangement in any given structure. This section is intended to provide a general overview of bar placement methods and sequences. As in all steel reinforcing work, the placing drawings will dictate the best sequence. They must be studied carefully so that an efficient plan of placing may be developed.

A beam is a horizontal structural member used to carry loads from a floor to columns, walls, and girders. A girder is a principal beam. The main difference between the two is that beams support other parts of a structure, while girders support beams. This difference becomes important when placing reinforcing steel in beams and girders because it affects the sequence of placing rebar.

In general, girders will be lower than the beams they support, so all bottom horizontal bars, truss bars, and stirrups must be placed in girders before any reinforcement is placed in beams. The actual placing procedures, however, are similar for both beams and girders. The placing of reinforcing steel in beams and girders begins after column vertical rebar are positioned and concrete has been placed to the bottom of the lowest beam or girder. Reinforcing steel is generally placed from bottom to top; that is, first in girders, then in beams, and finally in slabs and joists.

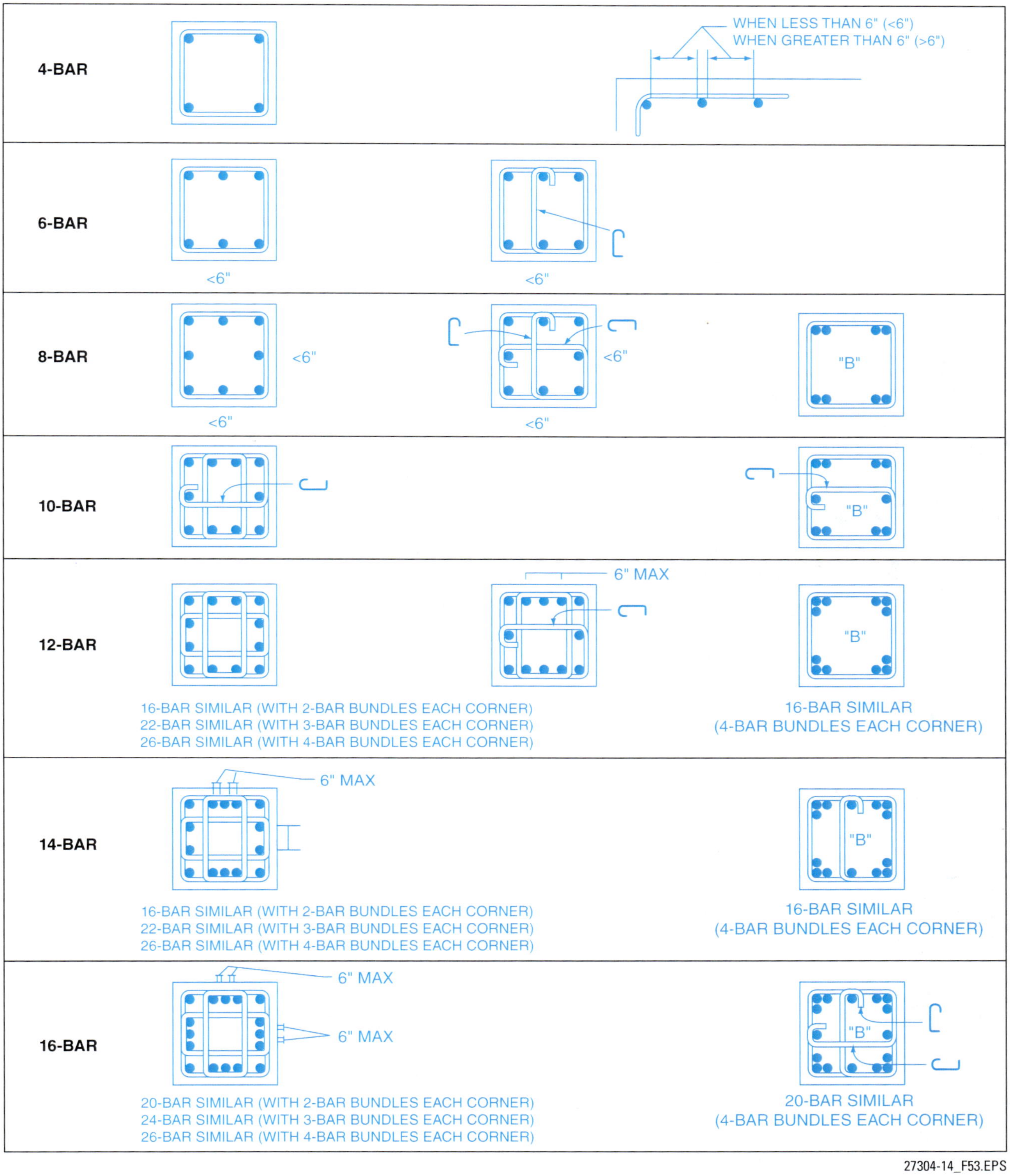

4-BAR

WHEN LESS THAN 6" (<6")
WHEN GREATER THAN 6" (>6")

6-BAR

<6"

<6"

8-BAR

<6"

<6"

<6"

<6"

"B"

10-BAR

"B"

12-BAR

6" MAX

"B"

16-BAR SIMILAR (WITH 2-BAR BUNDLES EACH CORNER)
22-BAR SIMILAR (WITH 3-BAR BUNDLES EACH CORNER)
26-BAR SIMILAR (WITH 4-BAR BUNDLES EACH CORNER)

16-BAR SIMILAR
(4-BAR BUNDLES EACH CORNER)

14-BAR

6" MAX

"B"

16-BAR SIMILAR (WITH 2-BAR BUNDLES EACH CORNER)
22-BAR SIMILAR (WITH 3-BAR BUNDLES EACH CORNER)
26-BAR SIMILAR (WITH 4-BAR BUNDLES EACH CORNER)

16-BAR SIMILAR
(4-BAR BUNDLES EACH CORNER)

16-BAR

6" MAX

6" MAX

"B"

20-BAR SIMILAR (WITH 2-BAR BUNDLES EACH CORNER)
24-BAR SIMILAR (WITH 3-BAR BUNDLES EACH CORNER)
26-BAR SIMILAR (WITH 4-BAR BUNDLES EACH CORNER)

20-BAR SIMILAR
(4-BAR BUNDLES EACH CORNER)

27304-14_F53.EPS

Figure 53 Standing column ties.

Placing Dowels

For dowels used in a continuous wall footing, the horizontal leg of the dowel is placed perpendicular to the length of the footing. Column or pier dowels placed in square, rectangular, or circular footings are placed such that the horizontal leg radiates from the center of the dowel configuration.

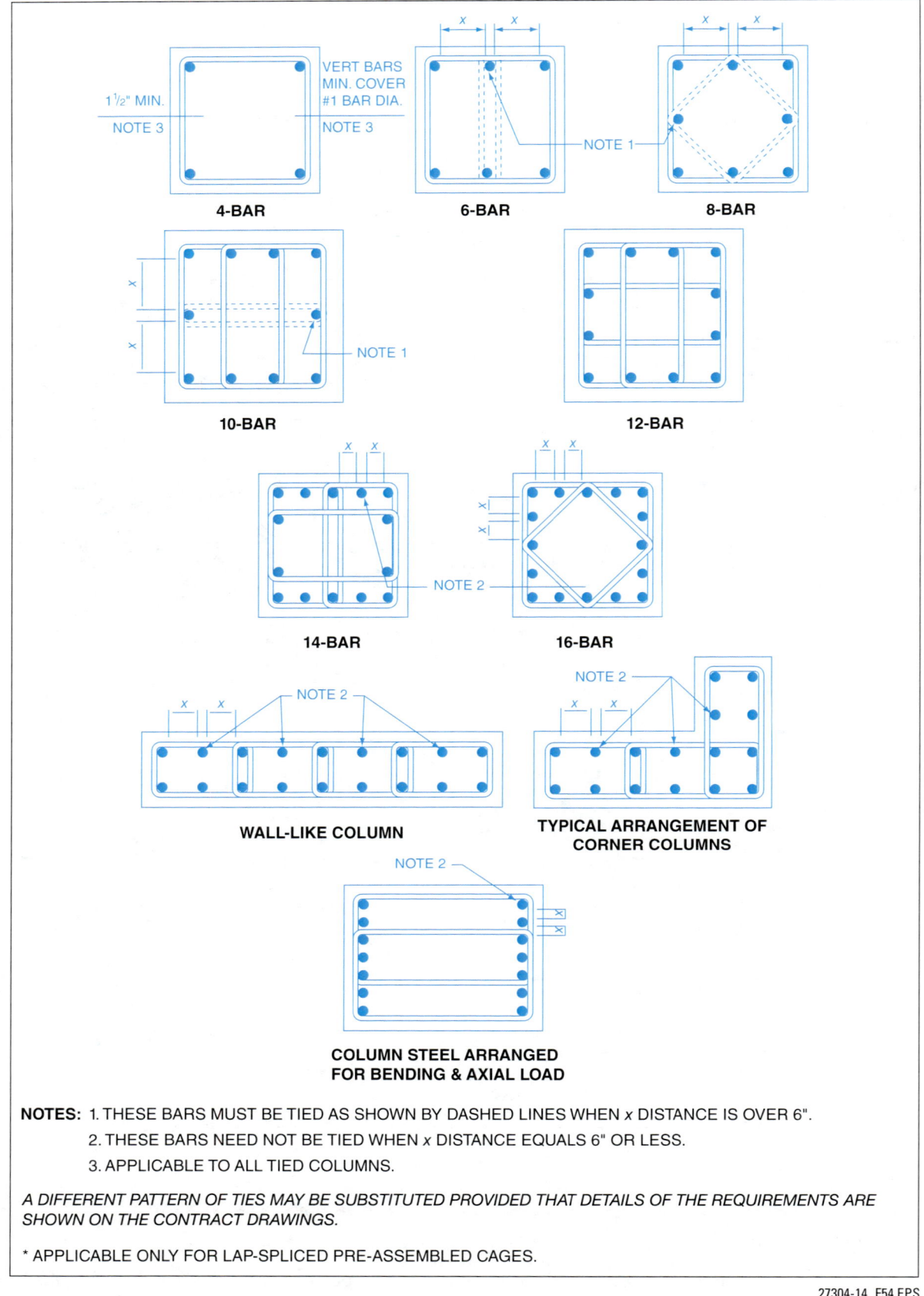

4-BAR **6-BAR** **8-BAR**

10-BAR **12-BAR**

14-BAR **16-BAR**

WALL-LIKE COLUMN **TYPICAL ARRANGEMENT OF CORNER COLUMNS**

COLUMN STEEL ARRANGED FOR BENDING & AXIAL LOAD

NOTES: 1. THESE BARS MUST BE TIED AS SHOWN BY DASHED LINES WHEN x DISTANCE IS OVER 6".

2. THESE BARS NEED NOT BE TIED WHEN x DISTANCE EQUALS 6" OR LESS.

3. APPLICABLE TO ALL TIED COLUMNS.

A DIFFERENT PATTERN OF TIES MAY BE SUBSTITUTED PROVIDED THAT DETAILS OF THE REQUIREMENTS ARE SHOWN ON THE CONTRACT DRAWINGS.

* APPLICABLE ONLY FOR LAP-SPLICED PRE-ASSEMBLED CAGES.

27304-14_F54.EPS

Figure 54 Closed column ties.

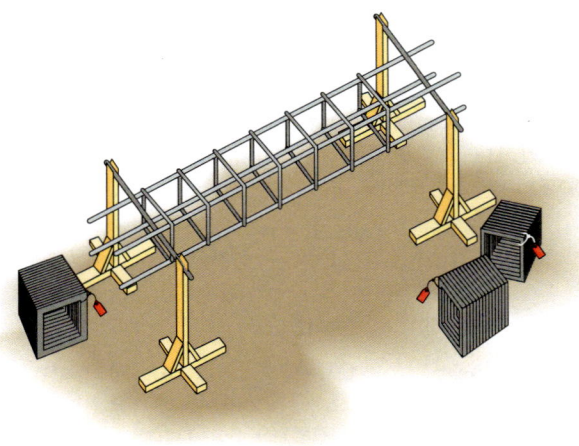

27304-14_F55.EPS

Figure 55 Rebar horse.

This sequence usually minimizes the need to manipulate rebar under one another at the points of intersection.

Figure 57 illustrates the general procedure for placing reinforcing steel in beams within the formwork. To place reinforcing steel in beams, proceed as follows:

Step 1 The beam bolsters are placed in the forms on centers not to exceed 5'.

Step 2 The stirrups and stirrup support bars are then placed in the forms. The forms should be marked with the proper spacing as found in the placing drawings.

Step 3 The straight bottom bars are placed next. In order to prevent these bars from moving during the concrete pour, they are often wired to the beam bolsters.

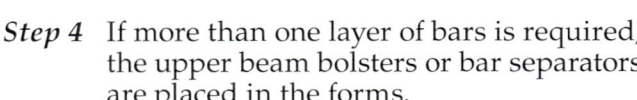

27304-14_F56.EPS

Figure 56 Spirals.

Step 4 If more than one layer of bars is required, the upper beam bolsters or bar separators are placed in the forms.

Step 5 The truss bars are placed last. A truss bar is a bar that has been bent in such a way that it serves as both top and bottom reinforcement. If truss bars are used in beams that require reinforcing placed in two layers, the truss bars should be placed directly over those in the lower layer, not in the spaces between the bars.

The reinforcing steel for beams and girders may also be pre-assembled. The sequence is usually as follows:

Step 1 Two straight bars are used as templates and marked with keel (marking crayon) or soapstone using the spacings found on the placing drawings.

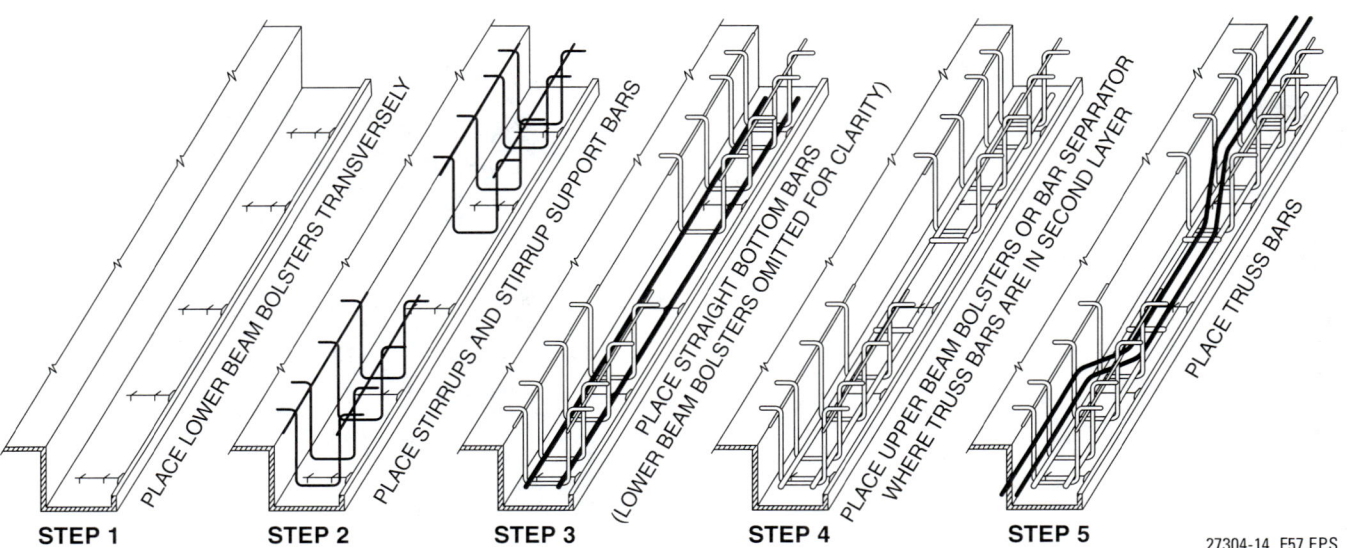

27304-14_F57.EPS

Figure 57 Beam placing sequence.

 27304-14 **Reinforcing Concrete**

Module Five 45

Step 2 The template bars are placed on rebar horses.

Step 3 Stirrups are placed on the marks and tied in place using saddle ties.

Step 4 A side bar is tied to the stirrups.

Step 5 At this point, the beam may be taken off the rebar horses and placed flat on the deck or ground. Diagonal wire braces may be tied to it to provide added rigidity.

Step 6 The bars may then be placed into the form and the bottom layers of reinforcing may be added.

Occasionally, the placing drawings will require closed ties instead of open, U-shaped stirrups. See *Figure 58*. Closed ties may be one piece with hooks in a corner or two pieces, known as cap ties, as shown in *Figure 59*.

The easier tie to use is, of course, the cap tie, because the bottom piece can be placed in the same way as a stirrup. After all other bars have been placed, the top piece of the cap tie may be placed and wired to the lower piece. If the drawings call for one-piece closed ties, these ties must be used. They can sometimes be slipped into place over the lengthwise bars. Usually, however, they must be sprung open enough so that they can be worked around the bars running lengthwise. This tends to twist the reinforcing steel out of shape and is also time consuming.

A joist is a small beam that is placed parallel between the main beams in floor construction. When joists are joined at the top to make a continuous structure, this structure is called a joist slab.

Joist slabs are used in situations where high loads are anticipated.

Placing reinforcing steel in joists begins when all the beam reinforcement has been placed. In general, the sequence for placing reinforcing steel in joists is as follows:

Step 1 Place joist chairs in the form, beginning 1" from the edge of each support. If possible, joist chairs should be spaced as close as 5', or as specified.

Step 2 Place the straight bottom bars on the joist chairs. If necessary, thread the bars between the beam stirrups and under the top beam bar. The bottom bars must extend into the supports at each end according to the specifications found in the placing drawings.

Step 3 Place the truss bars on the joist chairs next to the straight bottom bars. The bent-up ends of the truss bars must cross over all top bars in the beam and extend into the adjoining section by the amount required by the placing drawings. If no truss bars are required, two bottom straight bars are generally used. Likewise, one or two top bars are generally extended into the adjoining section. The extended ends of the truss bars or straight bars are supported on individual chairs placed on support bars. They may also rest on upper-joist chairs. The placing drawings will indicate the proper method of support.

Step 4 Place the distribution ribs next in one or two lines. These are also called continuous header joists or bridging ribs. They

Reinforcing-Steel Framework

This is an example of pre-assembled reinforcing-steel framework for a column that will eventually be raised into place. For ease of construction, rebar is often installed horizontally and then raised and placed into forms.

27304-14_SA07.EPS

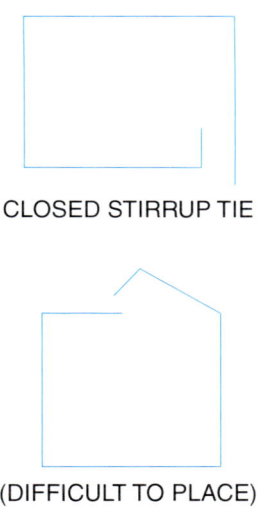

CLOSED STIRRUP TIE

(DIFFICULT TO PLACE)

27304-14_F58.EPS

Figure 58 Closed stirrup.

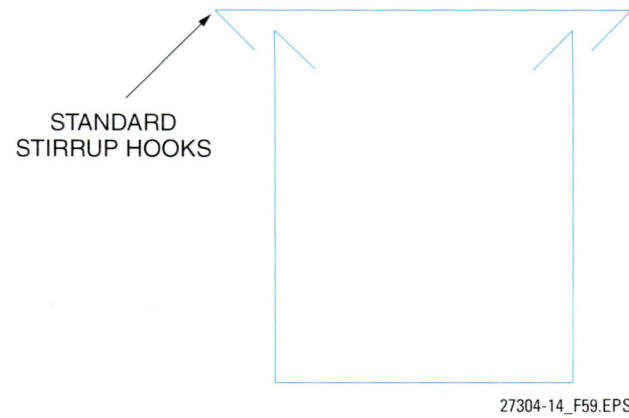

STANDARD
STIRRUP HOOKS

27304-14_F59.EPS

Figure 59 Capped stirrup.

extend the full length of the joist bay, are as deep as the joists, and are placed at right angles to the main joists. They provide lateral bracing for long spans of joists. Generally, there will be one rib placed at midspan for spans ranging from 18' to 24'. Bars in these ribs are usually shown on the floor joist plans.

Step 5 Place the **temperature bars** last. Their purpose is to minimize cracks due to changes in temperature and the normal shrinkage of concrete. Temperature bars are either #3 bars or welded-wire fabric reinforcement. If welded-wire fabric reinforcement is used, it should be unrolled so that it arches upward and then bends straight. Welded-wire fabric reinforcement is laid across the joists.

According to the direction of the main reinforcing run, slabs can be classified into two types. As its name indicates, a one-way slab contains reinforcement that runs in one direction between supports. A two-way slab contains reinforcement that runs in two directions between supports.

Reinforcement in one-way slabs consists of alternating straight bars and truss bars, or straight top and bottom bars running in one direction (*Figure 60*). Temperature bars are placed at right angles to the main reinforcing bars. Reinforcement is placed in one-way slabs in the same direction as the slab distributes the load applied to it.

The general procedure for placing and tying reinforcing steel in one-way slabs with straight bars and truss bars is as follows:

Step 1 Place slab bolsters so that they will lie at right angles to the main reinforcing. The

placing drawings provide the proper spacing.

Step 2 Place main reinforcing according to the placing drawings. Each group of steel is tied in place as it is set.

Step 3 Place high chairs at right angles to the main reinforcing.

Step 4 Place temperature bars at right angles to the main reinforcing.

Step 5 Place truss bars on high chairs. Ensure the truss bars are parallel with the main reinforcing.

Reinforcement is placed in a two-way slab in the same direction as the slab distributes the load applied to it. See *Figure 61*. The reinforcement usually consists of straight and truss bars or straight top and bottom bars arranged in **strips** called column strips and middle strips.

The width of each strip is generally one-half the distance between the centers of the columns. Column strips usually receive more reinforcement than middle strips. A proper placing se-

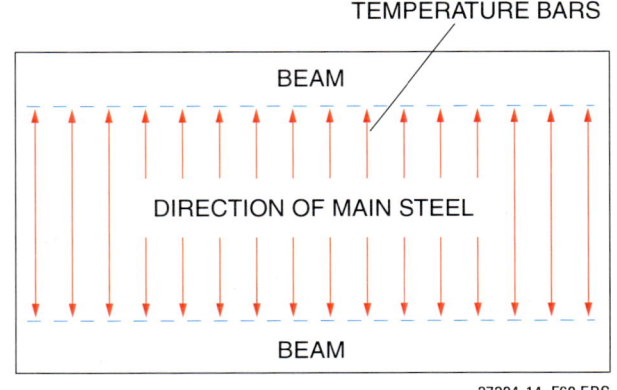

TEMPERATURE BARS

BEAM

DIRECTION OF MAIN STEEL

BEAM

27304-14_F60.EPS

Figure 60 One-way reinforced slab.

27304-14 *Reinforcing Concrete*

Module Five 47

quence must be followed to avoid threading the bars. Study the placing drawings very closely.

A general sequence for placing reinforcing steel in a two-way flat slab with straight and truss bars is as follows:

Step 1 Place continuous lines of slab bolsters in an east-west direction. Proper spacing is found on the placing drawings.

Step 2 Place the required lengths of slab bolsters in the east-west column strips at right angles to the slab.

Step 3 Place bottom straight bars running north-south in column and middle strips.

Step 4 Place bottom straight bars running east-west in column strips.

Step 5 Place three rows of #4 support bars on high chairs in an east-west direction at the head of each column. Tie the middle support bar to the column verticals.

Step 6 Place truss bars running north-south in column strips.

Step 7 Place straight top bars. These bars are usually placed within the bend-down point of the truss bars running east-west.

Step 8 Place truss bars running east-west in column strips. Usually, the east-west truss bars that rest upon the north-south, straight, middle-strip, and bottom bars

are tilted sideways so that they rest upon the top bars running north-south.

Step 9 Place three more rows of #4 support bars on high chairs in north-south and east-west column strips. Two rows should be placed at all slab edges.

Step 10 Place truss bars running north-south in the middle strip.

Reinforcing bars must be fabricated to conform to the designs drawn by the engineers for the particular structure. Most of the fabrication is done in shops, but a certain amount must always be done in the field. Fabrication of reinforcing bars includes the cutting, bending, and splicing of bars in accordance with certain tolerances and codes established by the American Concrete Institute.

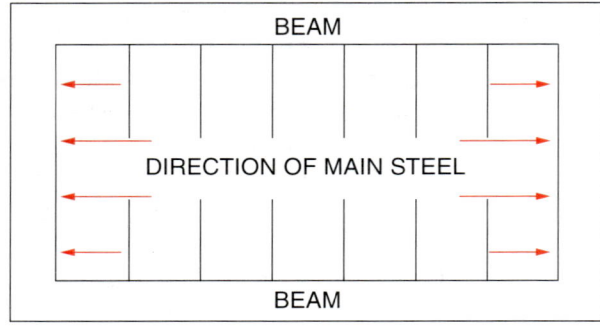

27304-14_F61.EPS

Figure 61 Two-way reinforced slab.

SUMMARY

Concrete used in structures must be properly reinforced with steel bars or welded-wire fabric reinforcement. Workers who place and tie rebar must be able to identify the various rebar types and sizes, as well as the various types of supports used to suspend rebar in the concrete.

In addition to describing concrete reinforcing materials and accessories, this module presented general information on cutting, bending, placing,

splicing, and tying reinforcing steel for reinforced concrete members. Even though there are general methods of placing reinforcing steel in various types of structural components, the placing drawings provide the exact specifications and locations governing the reinforcing steel. The most efficient system of bar placement should be determined from the placing drawings.

Review Questions

1. Unreinforced concrete typically can have a compressive strength of 2,000 to _____.
   a. 4,000 psi
   b. 5,000 psi
   c. 6,000 psi
   d. 7,000 psi

2. The adhesion of concrete to reinforcing steel is known as _____.
   a. chemical transformation
   b. the concrete bond
   c. the curing process
   d. sublimation

3. Column ties or column spirals are used with vertical reinforcement to prevent _____.
   a. buckling
   b. slump
   c. cracking
   d. compression

4. The supports at either end of a beam bridge are called abutments or _____.
   a. terminal piers
   b. anchor blocks
   c. end bents
   d. cantilevers

5. Strands of a posttensioning tendon are typically made from _____.
   a. copper
   b. aluminum
   c. steel
   d. brass

6. Placement of reinforcing bars may sometimes conflict with the position of other items buried in the slab (such as wiring conduits) that are referred to as _____.
   a. embedments
   b. submerged facilities
   c. mechanicals
   d. obstacles

7. Bundles of reinforcing bars should be hoisted using _____.
   a. nylon slings
   b. the bundle's wire wrapping
   c. heavy manila rope
   d. double chain slings

8. Expansion and contraction joints in highway construction can use short lengths of plain rebar called _____.
   a. pegs
   b. transfer pins
   c. dowels
   d. slip units

9. Rebar is manufactured to a standard length of _____.
   a. 45 feet
   b. 60 feet
   c. 75 feet
   d. 90 feet

10. Stainless steel reinforcing bars are typically used in _____.
    a. cantilevered structures
    b. high-temperature environments
    c. highly corrosive environments
    d. submerged structures

11. Each dimension of a standardized bar bend is designated by a _____.
    a. letter
    b. number
    c. letter/number combination
    d. symbol

12. The fabrication tolerance for straight reinforcing bars is a variation in length of _____.
    a. ½ inch
    b. ¾ inch
    c. 1 inch
    d. 1½ inches

13. Precast concrete blocks, which can be used as bar supports in footings, are also known as _____.

   a. darbies
   b. dillies
   c. derbies
   d. dobies

14. Plastic bar supports are used in situations where _____.

   a. #3 or smaller rebar is specified
   b. corrosion must be avoided
   c. cost is a factor
   d. ambient temperatures will be less than 80°F

15. Welded-wire fabric reinforcement is available in flat mats measuring _____.

   a. $2' \times 4'$
   b. $3' \times 6'$
   c. $4' \times 4'$
   d. $8' \times 16'$

16. Rebar with a diameter of ½" is designated as _____.

   a. #3
   b. #4
   c. #5
   d. #6

17. When a great deal of rebar cutting must be done on the job site, the cutting method of choice is _____.

   a. an oxyacetylene torch
   b. bolt cutters
   c. power shears
   d. a chop saw

18. The simplest type of rebar tie is the _____.

   a. snap tie
   b. figure-eight tie
   c. saddle tie
   d. wrap-and-snap tie

19. The recommended posture for tying horizontal, flat reinforcement is _____.

   a. squatting
   b. sitting
   c. kneeling
   d. standing stiff-legged

20. When making a spaced-lap splice, the center-to-center spacing of the bars must not exceed _____.

   a. 2"
   b. 4"
   c. 6"
   d. 8"

Trade Terms Quiz

Fill in the blank with the correct term that you learned from your study of this module.

1. The quantities, lengths, sizes, and grades of all bar materials to be used are shown on the _____.

2. Bars of a single length, size, or mark (bent) are often fastened together in a unit known as a(n) _____.

3. _____ are columns consisting of vertical reinforcing bars surrounded by a spiral that functions as a column tie.

4. Information on steel-reinforced concrete construction is available from _____.

5. When discussing a wall or other vertical concrete structure, the _____ is the one most distant from the person viewing it.

6. A(n) _____ is a hand tool used to provide leverage when manually bending bars or pipes.

7. _____ are anchoring devices embedded in concrete to receive bolts or screws that will attach shelf angles or other items, such as machinery.

8. The method of joining two reinforcing bars by placing them side by side and fastening them together is known as a(n) _____.

9. _____ is the common term for concrete that has been placed around metal bars or other reinforcing materials that improve its tensile and shear strength.

10. A noncontinuous beam that is supported at its two ends is referred to as a(n) _____.

11. The horizontal distance that a beam or girder stretches between supports is its _____.

12. Also called raiser bars, _____ rest upon individual high chairs to support the top bars in a slab.

13. _____ is twisted to fasten rebar together and hold it in the desired location until concrete is placed.

14. A(n) _____ is used to locate and hold dowels when laying out bolt holes and inserts in concrete construction.

15. In flat-slab or flat-plate construction, _____ are bands of reinforcing bars.

16. _____ are used in concrete to minimize shrinkage and cracks due to temperature changes.

17. A reinforced wall used to hold back soil or materials such as coal or sand is referred to as a(n) _____.

18. A(n) _____ is a concrete slab with drop panels that is reinforced in two directions.

19. Placed atop a pile, a(n) _____ is used to distribute loads from a structure to the supporting pile.

20. Two adjacent separately cast sections of concrete are connected using metal bars called _____.

21. The distance from the topmost layer of reinforcing steel to the face of the concrete is called by various names, but _____ is most common.

22. _____ are cast-in-place, drilled hole piles. The term also refers to piers extending down to solid earth or rock through a layer of soft soil or water.

23. The support structure located at each end of a bridge is a(n) _____.

24. A(n) _____ is a frame with two or more legs that is placed perpendicular (at a right angle) to the length of the structure it supports.

25. A method of reinforcing-bar connection that involves lapping the bars in direct contact is called a(n) _____.

26. When discussing a wall or other vertical concrete structure, the _____ is the one closest to the person viewing it.

27. In a spiral, the _____ is the center-to-center spacing between the turns.

 27304-14 Reinforcing Concrete

28. Bars, dowels, or anchor bolts may be enclosed by tubes known as _____.

29. Bars that are spliced at different points are said to be joined with _____.

30. Moisture is drained from the interior of a wall by passing through a(n) _____.

31. A(n) _____ is a construction document that lists, in table form, similar items.

32. The principal beam that supports other beams is called a(n) _____.

33. _____ are support devices, usually used in pairs, to hold reinforcement at a comfortable level for fastening while prefabricating columns.

34. A beam that extends over at least three supports (including the end supports) is classified as a(n) _____.

35. Also known as column ties, _____ are reinforcing steel wrapped around the vertical bars serving as column reinforcement.

36. _____ are steel bars wrapped around the vertical reinforcement of a column to prevent buckling under compression load.

37. A concrete wall with a layer of reinforcement for each face is a(n) _____.

38. Information on reinforcing-bar size, location, spacing and other needed data is contained in the _____.

39. _____ are U-shaped or box-shaped reinforcing bars placed perpendicular to the longitudinal bars in beams to improve shear strength.

40. _____ are semicircular or right-angle bends on the free end of a reinforcing bar to anchor it in concrete.

41. A horizontal structural member is classified as a(n) _____.

42. A(n) _____ is a structural member carrying a primarily vertical load as it supports a beam or other horizontal member.

43. A reinforcing bar that is bent into a box shape, a(n) _____ holds together the longitudinal reinforcing bars in a beam or column.

44. In a(n) _____, horizontal or vertical reinforcing bars are arranged in a single layer positioned in the center of the wall.

Trade Terms

Abutment	Column ties	Girder	Rebar horses	Stirrups
Band	Concrete cover	Hickey bar	Reinforced concrete	Strips
Bar list	Contact splice	Hook	Retaining wall	Support bars
Beam	Continuous beam	Inserts	Schedule	Temperature bars
Bent	CRSI	Lapped splice	Simple beam	Template
Bundle of bars	Double-curtain wall	Near face	Single-curtain wall	Tie
Caissons	Dowel	Pile cap	Sleeve	Tie wire
Column	Far face	Pitch	Span	Weephole
Column spirals	Flat slab	Placing drawings	Staggered splices	

Trade Terms Introduced in This Module

Abutment: The supporting substructure at each end of a bridge.

Band: Reinforcing steel in columns that is wrapped around the vertical bars to counteract compression forces.

Bar list: A bill of materials for a job site that shows all bar quantities, sizes, lengths, grades, placement areas, and bending dimensions to be used.

Beam: A horizontal structural member.

Bent: A self-supporting frame having at least two legs and placed at right angles to the length of the structure it supports, such as the columns and cap supporting the spans of a bridge.

Bundle of bars: A bundle consisting of one size, length, or mark (bent) of bar, with the following exceptions: very small quantities may be bundled together for convenience, and groups of varying bar lengths or marks that will be placed adjacent to one another may be bundled together.

Caissons: Piers usually extending through water or soft soil to solid earth or rock; also refers to cast-in-place, drilled-hole piles.

Column: A post or vertical structural member supporting a floor beam, girder, or other horizontal member and carrying a primarily vertical load.

Column spirals: Columns in which the vertical bars are enclosed within a spiral that functions like a column tie.

Column ties: Bars that are bent into square, rectangular, U-shaped, circular, or other shapes for the purpose of holding vertical column bars laterally in place and that prevent buckling of the vertical bars under compression load.

Concrete cover: The distance from the face of the concrete to the reinforcing steel; also referred to as fireproofing, clearance, or concrete protection.

Contact splice: A means of connecting reinforcing bars by lapping in direct contact.

Continuous beam: A beam that extends over three or more supports (including end supports).

CRSI: Concrete Reinforcing Steel Institute.

Double-curtain wall: A concrete wall that contains a layer of reinforcement at each face.

Dowel: A bar connecting two separately cast sections of concrete. A bar extending from one concrete section into another is said to be doweled into the adjoining section.

Far face: The face farthest from the viewer (as of a wall); may be the outside or inside face, depending on whether one is inside looking out or outside looking in.

Flat slab: A concrete slab reinforced in two or more directions, with drop panels but generally without beams, and with or without column capitals.

Girder: The principal beam supporting other beams.

Hickey bar: A hand tool with a side-opening jaw used in developing leverage for making in-place bends on bars or pipes.

Hook: A 180-degree (semicircular) or 90-degree turn at the free end of a bar to provide anchorage in concrete. For stirrups and column ties only, turns of either 90 degrees or 135 degrees are used.

Inserts: Devices that are positioned in concrete to receive a bolt or screw to support shelf angles, machinery, etc.

Lapped splice: The joining of two reinforcing bars by lapping them side by side, or the length of overlap of two bars; similarly, the side and end overlap of sheets or rolls of welded-wire fabric reinforcement.

Near face: The face nearest the viewer, which may be inside or outside, depending on whether one is inside looking out or outside looking in.

Pile cap: A structural member placed on the tops of piles and used to distribute loads from the structure to the piles.

Pitch: The center-to-center spacing between the turns of a spiral.

Placing drawings: Detailed drawings that give the bar size, location, spacing, and all other information required to place the reinforcing steel.

Rebar horses: Wood or metal supports that are used in groups of two or more to hold main reinforcing in a convenient position for placing ties while prefabricating column, beam, or pile cages.

Reinforced concrete: Concrete that has been placed around some type of steel reinforcement material. After the concrete cures, the reinforcement provides greater tensile and shear strength for the concrete. Almost all concrete is reinforced in some manner.

Retaining wall: A wall that has been reinforced to hold or retain soil, water, grain, coal, or sand.

Schedule: A table on placing drawings that lists the size, shape, and number of bars each way, and the mark number of the bars if they are bent.

Simple beam: A beam supported at each end (two points) and not continuous.

Single-curtain wall: A concrete wall that contains a single layer of vertical or horizontal reinforcing bars in the center of the wall.

Sleeve: A tube that encloses a bar, dowel, anchor bolt, or similar item.

Span: The horizontal distance between supports of a member such as a beam, girder, slab, or joist; also, the distance between the piers or abutments of a bridge.

Staggered splices: Splices in bars that are not made at the same point.

Stirrups: Reinforcing bars used in beams for shear reinforcement; typically bent into a U shape or box shape and placed perpendicular to the longitudinal steel.

Strips: Bands of reinforcing bars in flat-slab or flat-plate construction. The column strip is a quarter-panel wide on each side of the column center line and runs from column to column. The middle strip is half a panel in width, filling in between column strips, and runs parallel to the column strips.

Support bars: Bars that rest upon individual high chairs or bar chairs to support top bars in slabs or joists, respectively. They are usually #4 bars and may replace a like number of temperature bars in slabs when properly lap spliced; also used longitudinally in beams to provide support for the tops of stirrups. Also called raiser bars.

Temperature bars: Bars distributed throughout the concrete to minimize cracks due to temperature changes and concrete shrinkage.

Template: A device used to locate and hold dowels, to lay out bolt holes and inserts, etc.

Tie: A reinforcing bar bent into a box shape and used to hold longitudinal bars together in columns and beams. Also known as stirrup ties.

Tie wire: Wire (generally #16, #15, or #14 gauge) used to secure rebar intersections for the purpose of holding them in place until concreting is completed.

Weephole: A drainage opening in a wall.

Additional Resources

This module presents thorough resources for task training. The following resource material is suggested for further study.

29 CFR 1926, *Safety and Health Regulations for Construction*, Latest Edition. Washington, D.C.: Occupational Safety and Health Administration.

ACI 315, Details and Detailing of Concrete Reinforcement, Latest Edition. Farmington Hills, MI: American Concrete Institute.

ACI 318-95, Building Code Requirements for Structural Concrete, Latest Edition. Farmington Hills, MI: American Concrete Institute.

ASTM A615, Standard Specification for Deformed and Plain Carbon-Steel Bars for Concrete Reinforcement, Latest Edition. West Conshohocken, PA: ASTM International.

ASTM A706, Standard Specification for Low-Alloy Steel Deformed Bars and Plain Bars for Concrete Reinforcement, Latest Edition. West Conshohocken, PA: ASTM International.

ASTM A996, Standard Specification for Rail-Steel and Axle-Steel Deformed Bars for Concrete Reinforcement, Latest Edition. West Conshohocken, PA: ASTM International.

Manual of Standard Practice, Latest Edition. Concrete Reinforcing Steel Institute (CRSI).

Placing Reinforcing Bars. 2005. Concrete Reinforcing Steel Institute (CRSI).

Figure Credits

Courtesy of Portland Cement Association, CO01, Figure 13, Figures 15–16, SA02

www.rebar.net, Figure 21, Figure 22

Courtesy of JET Tools, Figure 29

Benner-Nawman, Inc., SA04

BN Products – USA, Figure 31

Klein Tools, Inc., Figure 39

Bar Splice Products, Inc., SA05

MAX USA Corp., SA06

Haskell, Figure 50b

Brice Building Company, SA07

Section Review Answers

Answer	Section Reference	Objective Reference
Section One		
1. d	1.1.0	1a
2. b	1.2.2	1b
3. True	1.3.0	1c
Section Two		
1. d	2.0.0	2
2. a	2.1.0	2a
3. b	2.2.0	2b
4. d	2.4.3	2d
5. a	2.5.2	2e
Section Three		
1. False	3.1.0	3a
2. c	3.2.0	3b
Section Four		
1. c	4.1.0	4a
2. c	4.2.1	4b

NCCER CURRICULA — USER UPDATE

NCCER makes every effort to keep its textbooks up-to-date and free of technical errors. We appreciate your help in this process. If you find an error, a typographical mistake, or an inaccuracy in NCCER's curricula, please fill out this form (or a photocopy), or complete the online form at **www.nccer.org/olf**. Be sure to include the exact module ID number, page number, a detailed description, and your recommended correction. Your input will be brought to the attention of the Authoring Team. Thank you for your assistance.

Instructors – If you have an idea for improving this textbook, or have found that additional materials were necessary to teach this module effectively, please let us know so that we may present your suggestions to the Authoring Team.

NCCER Product Development and Revision
13614 Progress Blvd., Alachua, FL 32615

Email: curriculum@nccer.org
Online: www.nccer.org/olf

❏ Trainee Guide   ❏ Lesson Plans   ❏ Exam   ❏ PowerPoints   Other _____

Craft / Level: _____ Copyright Date: _____

Module ID Number / Title: _____

Section Number(s): _____

Description: _____

_____

_____

_____

Recommended Correction: _____

_____

_____

_____

Your Name: _____

Address: _____

Email: _____ Phone: _____